AXIOMATIC
SET THEORY

AXIOMATIC

SET THEORY

by

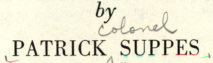

PATRICK SUPPES

Professor of Philosophy and Statistics
Stanford University

DOVER PUBLICATIONS, INC.

NEW YORK

Published in Canada by General Publishing Company, Ltd., 30 Lesmill Road, Don Mills, Toronto, Ontario.
Published in the United Kingdom by Constable and Company, Ltd., 10 Orange Street, London WC 2.

This Dover edition, first published in 1972, is an unabridged and corrected republication of the work originally published by D. Van Nostrand Company in 1960, to which the author has added a new preface and a new section (8.4).

International Standard Book Number: 0-486-61630-4
Library of Congress Catalog Card Number: 72-86226

Manufactured in the United States of America
Dover Publications, Inc.
180 Varick Street
New York, N.Y. 10014

To
MY PARENTS

PREFACE TO THE DOVER EDITION

The present edition differs from the first edition in the correction of all known errors, and in the addition of a final section (8.4) on the independence of the axiom of choice and the generalized continuum hypothesis. For bringing to my attention numerous misprints and minor mistakes I am indebted to a great many people, in fact, too many to list here; but I do want to mention Nuel Belnap, Theodore Hailperin, Craig Harrison, Milton Levy, Elliott Mendelson, N. C. K. Phillips, D. H. Potts, Hilary Staniland, and John Wallace.

PATRICK SUPPES

Stanford, California
June, 1972

PREFACE TO THE FIRST EDITION

This book is intended primarily to serve as a textbook for courses in axiomatic set theory. The Zermelo-Fraenkel system is developed in detail. The mathematical prerequisites are minimal; in particular, no previous knowledge of set theory or mathematical logic is assumed. On the other hand, students will need a certain degree of general mathematical sophistication, especially to master the last two chapters. Although some logical notation is used throughout the book, proofs are written in an informal style and an attempt has been made to avoid excessive symbolism. A glossary of the more frequently used symbols is provided.

The eight chapters are organized as follows. Chapter 1 provides a brief introduction. Chapter 2 is concerned with general developments, and Chapter 3 with relations and functions. There is not a single difficult theorem in these first three chapters, and they can be covered very rapidly in an advanced class. The main pedagogical emphasis is on the exact role of the various axioms introduced in Chapter 2, which are summarized at the end of the chapter.

Chapter 4 considers the classical topics of equipollence of sets, finite sets and cardinal numbers. The Schröder-Bernstein Theorem is proved early in the chapter. The development of the theory of finite sets follows closely Alfred Tarski's well known article of 1924. The theory of cardinal numbers is simplified by introducing a special axiom to the effect that the cardinal numbers of two equipollent sets are identical. This axiom is not part of the standard Zermelo-Fraenkel system, and so every definition or theorem which depends on it has been marked by the symbol '†', but it leads to such a simple and natural development that I feel its introduction is fully justified. Chapter 5 covers some of the same ground from another viewpoint. The natural numbers are defined as von Neumann ordinals and the theory of recursive definitions is developed. The axiom of infinity is introduced, and the final section covers the basic facts about denumerable sets.

In Chapter 6 the standard construction of the rational and real numbers is given in detail. Cauchy sequences of rational numbers rather than Dedekind cuts are used to define the real numbers. Most of the elementary facts about sets of the power of the continuum are proved in the final section.

Because for many courses in set theory it will not be feasible in the time allotted to include the construction of the real numbers or because the topic may be assigned to other courses, the book has been written so that this chapter may be omitted without loss of continuity.

Chapter 7 is concerned with transfinite induction and ordinal arithmetic. The treatment of transfinite induction and definition by transfinite recursion is one of the most detailed in print. Numerous variant formulations have been given in the hope that successive consideration of them will clarify for the student the essential character of transfinite processes. The axiom schema of replacement is introduced in connection with establishing an appropriate recursion schema to define ordinal addition. The more familiar facts about alephs and well-ordered sets are proved in the latter part of the chapter.

Chapter 8 deals mainly with the axiom of choice and its equivalents, like Hausdorff's Maximal Principle and Zorn's Lemma. Important facts whose known proofs require the axiom of choice are also deferred consideration until this chapter. A typical example is the identity of ordinary and Dedekind infinity.

Although the list of references at the end of the book is small compared to that given in Fraenkel's *Abstract Set Theory*, I have attempted to refer to many of the more important papers for the topics considered. Because set theory, even perhaps axiomatic set theory, is finally coming to be a staple of every young mathematician's intellectual fare, it is of some historical interest to note that the majority of the papers referred to in the text were published before 1930.

I hope that this book will prove useful in connection with several kinds of courses. A semester mathematics course in set theory for seniors or first-year graduate students should be able to cover the whole book with the exception perhaps of Chapter 6. Philosophy courses in the foundations of mathematics may profitably cover only the first four chapters, which end with the construction of the natural numbers as finite cardinals. The material in the first six chapters, ending with the construction of the real numbers, is suitable for an undergraduate mathematics course in the foundations of analysis, or as auxiliary reading for the course in the theory of functions of a real variable.

This book was begun in 1954 as a set of lecture notes. I have revised the original version at least four times and in the process have benefited greatly from the criticisms and remarks of many people. I want particularly to thank Michael Dummett, John W. Gray, Robert McNaughton, John Myhill, Raphael M. Robinson, Richard Robinson, Herman and Jean Rubin, Dana Scott, J. F. Thomson and Karol Valpreda Walsh. During the academic year 1957-58 Joseph Ullian used a mimeographed version at Stanford, and most of his many useful criticisms and corrections have been incorporated.

As in the case of my *Introduction to Logic*, I owe a debt of gratitude too large to measure to Robert Vaught for his detailed and penetrating comments on the next-to-final draft. David Lipsich read the galley proofs and suggested several desirable changes most of which could be accommodated. Mrs. Louise Thursby has typed with accuracy and patience more versions of the text than I care to remember. Mrs. Blair McKnight has been of much assistance in reading proofs and preparing the indexes. Any errors which remain are my sole responsibility.

PATRICK SUPPES

Stanford, California
January, 1960

TABLE OF CONTENTS

CHAPTER 1

INTRODUCTION

§ 1.1 Set Theory and the Foundations of Mathematics. Among the many branches of modern mathematics set theory occupies a unique place: with a few rare exceptions the entities which are studied and analyzed in mathematics may be regarded as certain particular sets or classes of objects.* This means that the various branches of mathematics may be formally defined within set theory. As a consequence, many fundamental questions about the nature of mathematics may be reduced to questions about set theory.

The working mathematician, as well as the man in the street, is seldom concerned with the unusual question: What is a number? But the attempt to answer this question precisely has motivated much of the work by mathematicians and philosophers in the foundations of mathematics during the past hundred years. Characterization of the integers, rational numbers and real numbers has been a central problem for the classical researches of Weierstrass, Dedekind, Kronecker, Frege, Peano, Russell, Whitehead, Brouwer, and others. Perplexities about the nature of number did not originate in the nineteenth century. One of the most magnificent contributions of ancient Greek mathematics was Eudoxus' theory of proportion, expounded in Book V of Euclid's *Elements;* the main aim of Eudoxus was to give a rigorous treatment of irrational quantities like the geometric mean of 1 and 2. It may indeed be said that the detailed development from the general axioms of set theory of number theory and analysis is very much in the spirit of Eudoxus.

Yet the real development of set theory was not generated directly by an attempt to answer this central problem of the nature of number, but by the researches of Georg Cantor around 1870 in the theory of infinite series

---

*Intuitively we mean by *set* or *class* a collection of entities of any sort. Thus we can speak of the set of all Irishmen, or the set of all prime numbers. In ordinary mathematics the words 'set', 'class', 'collection', and 'aggregate' are synonyms, and the same shall be true here, except in a few explicitly noted contexts.

and related topics of analysis.* Cantor, who is usually considered the founder of set theory as a mathematical discipline, was led by his work into a consideration of infinite sets or classes of arbitrary character. In 1874 he published his famous proof that the set of real numbers cannot be put into one-one correspondence with the set of natural numbers (the non-negative integers). In 1878 he introduced the fundamental notion of two sets being equipollent or having the same power (*Mächtigkeit*) if they can be put into one-one correspondence with each other. Clearly two finite sets have the same power just when they have the same number of members. Thus the notion of power leads in the case of infinite sets to a generalization of the notion of a natural number to that of an infinite cardinal number. Development of the general theory of transfinite numbers was one of the great accomplishments of Cantor's mathematical researches.

Technical consideration of the many basic concepts of set theory introduced by Cantor will be given in due course. From the standpoint of the foundations of mathematics the philosophically revolutionary aspect of Cantor's work was his bold insistence on the actual infinite, that is, on the existence of infinite sets as mathematical objects on a par with numbers and finite sets. Historically the concept of infinity has played a role in the literature of the foundations of mathematics as important as that of the concept of number. There is scarcely a serious philosopher of mathematics since Aristotle who has not been much exercised about this difficult concept.

Any book on set theory is naturally expected to provide an exact analysis of the concepts of number and infinity. But other topics, some controversial and important in foundations research, are also a traditional part of the subject and are consequently treated in the chapters that follow. Typical are algebra of sets, general theory of relations, ordering relations in particular, functions, finite sets, cardinal numbers, infinite sets, ordinal arithmetic, transfinite induction, definition by transfinite recursion, axiom of choice, Zorn's Lemma. At this point the reader is not expected to know what these phrases mean, but such a list may still give a clue to the more detailed contents of this book.

In this book set theory is developed axiomatically rather than intuitively. Several considerations have guided the choice of an axiomatic approach. One is the author's opinion that the axiomatic development of set theory is among the most impressive accomplishments of modern mathematics. Concepts which were vague and unpleasantly inexact for decades and sometimes even centuries can be given a precise meaning. Adequate axioms for set theory provide one clear, constructive answer to the question: Exactly what assumptions, beyond those of elementary logic, are required

---

*For a detailed historical survey of Cantor's work, see Jourdain's *Introduction to Cantor* [1915].

as a basis for modern mathematics? The most pressing consideration, however, is the discovery, made around 1900, of various paradoxes in naive, intuitive set theory, which admits the existence of sets of objects having any definite property whatsoever. Some particular restricted axiomatic approach is needed to avoid these paradoxes, which are discussed in §§ 1.3 and 1.4 below.

§ **1.2 Logic and Notation.** We shall use symbols of logic extensively for purposes of precision and brevity, particularly in the early chapters. But proofs are mainly written in an informal style. The theory developed is treated as an axiomatic theory of the sort familiar from geometry and other parts of mathematics, and not as a formal logistic system for which exact rules of syntax and semantics are given. The explicitness of proofs is sufficient to make it a routine matter for any reader familiar with mathematical logic to provide formalized proofs in some standard system of logic. However, familiarity with mathematical logic is not required for understanding any part of the book.

At this point we introduce the few logical symbols which will be used. We first consider five symbols for the five most common sentential connectives. The negation of a formula P is written as $-P$. The conjunction of two formulas P and Q is written as $P \& Q$. The disjunction of P and Q as $P \vee Q$. The implication with P as antecedent and Q as consequent as $P \to Q$. The equivalence P *if and only if* Q as $P \leftrightarrow Q$. The universal quantifier *For every* v as $(\forall v)$, and the existential quantifier *For some* v as $(\exists v)$. We also use the symbol $(E!v)$ for *There is exactly one* v *such that*. This notation may be summarized in the following table.

LOGICAL NOTATION

$-P$	It is not the case that P
$P \& Q$	P and Q
$P \vee Q$	P or Q
$P \to Q$	If P then Q
$P \leftrightarrow Q$	P if and only if Q
$(\forall v)P$	For every v, P
$(\exists v)P$	For some v, P
$(E!v)P$	There is exactly one v such that P

Thus the sentence:

For every x there is a y such that $x < y$

is symbolized:

(1) $$(\forall x)(\exists y)(x < y).$$

The sentence:

> For every ϵ there is a δ such that for every y
> if $|x - y| < \delta$ then $|f(x) - f(y)| < \epsilon$

is symbolized:

$$(2) \qquad (\forall \epsilon)(\exists \delta)(\forall y)(|x - y| < \delta \rightarrow |f(x) - f(y)| < \epsilon).$$

The sentence:

> For every x there is exactly one y such that $x + y = 0$

is symbolized:

$$(\forall x)(E!y)(x + y = 0).$$

A given logical symbol may correspond to several English idioms. Thus $(\forall v)P$ may be read *For all* v, P as well as *For every* v, P. Sentences (1) and (2) illustrate the use of parentheses for purposes of punctuation. No formal explanation seems necessary. However, one convention concerning the relative dominance of the sentential connectives &, $\vee$, $\rightarrow$ and $\leftrightarrow$ will reduce considerably the number of parentheses. The convention is that $\leftrightarrow$ and $\rightarrow$ dominate & and $\vee$. Thus, the formula:

$$(x < y \,\&\, y < z) \rightarrow x < z$$

may be written without parentheses:

$$(3) \qquad x < y \,\&\, y < z \rightarrow x < z.$$

Similarly,

$$x + y \neq 0 \leftrightarrow (x \neq 0 \vee y \neq 0)$$

may be written:

$$x + y \neq 0 \leftrightarrow x \neq 0 \vee y \neq 0.$$

Principles of logic which are needed in the sequel and which may not be familiar to some readers will be intuitively explained when used. One principle used, concerning which there is some disagreement in practice among mathematicians, is that the double bar '$=$' is taken as the sign of identity. The formula '$x = y$' may be read 'x is the same as y', 'x is identical with y', or 'x is equal to y'. The last reading is permissible here only if it is understood that equality means sameness of identity (which is what it does mean in almost all ordinary mathematical contexts). The exact status of the relation of identity within set theory is discussed in §2.2.

A few remarks concerning quantifiers may also be helpful. The *scope* of a quantifier is the quantifier itself together with the smallest formula immediately following the quantifier. What the smallest formula is, is always indicated by parentheses. Thus in the formula

$$(4) \qquad (\exists x)(x < y) \vee y = 0$$

the scope of the quantifier '$(\exists x)$' is the formula '$(\exists x)(x < y)$'. Following an almost universal practice in mathematics, we shall omit, in the formulation of *axioms and theorems*, any universal quantifier whose scope is the whole formula. For instance, instead of (1) above, we would write: $(\exists y)$ $(x < y)$.

In a few places we shall need the notions of *bound* and *free* variables. An occurrence of a variable in a formula is bound if and only if this occurrence is within the scope of a quantifier using this variable. An occurrence of a variable in a formula is free if not bound. Finally, a variable is a *bound variable* in a formula if and only if at least one occurrence is bound; it is a free variable in a formula if and only if at least one occurrence is free. In formula (1) of this section all variables are bound; in (3) all variables are free; in (4) 'x' is bound and 'y' is free. By virtue of the convention stated in the preceding paragraph concerning omission of universal quantifiers in axioms and theorems, all variables occurring in axioms and theorems are bound.

§ **1.3 Axiom Schema of Abstraction and Russell's Paradox.** In his initial development of set theory, Cantor did not work explicitly from axioms. However, analysis of his proofs indicates that almost all of the theorems proved by him can be derived from three axioms: (i) The axiom of extensionality for sets, which asserts that two sets are identical if they have the same members; (ii) the axiom of abstraction, which states that given any property there exists a set whose members are just those entities having that property; (iii) the axiom of choice, which will not be formulated at this point and is not pertinent to our discussion of the paradoxes.

The source of trouble is the axiom of abstraction. The first explicit formulation of it seems to be as Axiom V in Frege [1893]. In 1901 Bertrand Russell discovered that a contradiction could be derived from this axiom by considering the set of all things which have the property of not being members of themselves.* Because this paradox was historically important

---

*Frege stated his own reaction to Russell's paradox in a famous appendix to the second volume of his *Grundgesetze der Arithmetik*, published in 1903. The translation of the opening lines given here is from Geach and Black [1952, p. 234].

"Hardly anything more unfortunate can befall a scientific writer than to have one of the foundations of his edifice shaken after the work is finished.

"This was the position I was placed in by a letter of Mr. Bertrand Russell, just when the printing of this volume was nearing its completion. It is a matter of my Axiom (V). I have never disguised from myself its lack of the self-evidence that belongs to the other axioms and that must properly be demanded of a logical law . . . I should gladly have dispensed with this foundation if I had known of any substitute for it. And even now I do not see how arithmetic can be scientifically established; how numbers can be apprehended as logical objects, and brought under review; unless we are permitted — at least conditionally — to pass from a concept to its extension. May I always speak of the extension of a concept — speak of a class? And if not, how are the exceptional cases recognized? . . . These are the questions raised by Mr. Russell's communication." For a recent discussion of Frege's appendix, see Quine [1955].

in motivating the development of new, restricted axioms for set theory, its derivation will be given here. For symbolic formulation we need to introduce the binary predicate '$\in$' of set membership. The formula '$x \in y$' is read 'x is a member of y', 'x belongs to y', or sometimes, 'x is in y'. Thus, if A is the set of first five odd positive integers, the sentence '$7 \in A$' is true and '$6 \in A$' is false.

Using '$\in$' and the logical notation introduced in the previous section, we may give a precise formulation of the axiom of abstraction:

(1) $(\exists y)(\forall x)(x \in y \leftrightarrow \varphi(x))$,

where it is understood that $\varphi(x)$ is a formula in which the variable 'y' is not free. To obtain Russell's paradox, we want $\varphi(x)$ to assert that x is not a member of itself. The appropriate formula is clearly:

$$-(x \in x).$$

We then have as an instance of the axiom of abstraction:

(2) $(\exists y)(\forall x)(x \in y \leftrightarrow -(x \in x))$.

Taking $x = y$ in (2), we infer:

(3) $y \in y \leftrightarrow -(y \in y)$,

which is logically equivalent to the contradiction:

(4) $y \in y \ \& \ -(y \in y)$.

This simple derivation has far-reaching consequences for the axiomatic foundations of set theory. It plainly shows that in admitting (1) as an axiom we have granted too much. If we adhere to ordinary logic we cannot in a self-consistent manner claim that for every property there is a corresponding *set* of things having that property.

In considering how to build anew the foundations of set theory, perhaps the first thing to notice is that the axiom of abstraction is really an infinite bundle of axioms rather than a single axiom: when we replace the expression '$\varphi(x)$' in (1) by *any* formula in which 'y' is not a free variable, we have a new axiom. An axiom which permits this sort of blanket substitution of formulas is usually called an *axiom schema*. The reason for using the word 'schema' should be obvious. As it stands (1) is not a definite assertion, but a scheme for making many assertions. From the schema we obtain a definite assertion by substituting a definite formula for '$\varphi(x)$'.

The axiom schema which we shall use is due to Ernst Zermelo [1908], and is usually called the *axiom schema of separation* (Aussonderung Axiom) because it permits us to separate off the elements of a given set which satisfy some property and form the set consisting of just these elements. Thus if we know that the set of animals exists, we may use the axiom *schema* of separation to assert the existence of the set of animals which have the property of being men. That is, the property of being human enables us to separate men from other animals. Corresponding to (1) the precise form of the axiom is:

(5) $$(\exists y)(\forall x)[x \in y \leftrightarrow x \in z \;\&\; \varphi(x)].$$

The change from (1) to (5) is slight but potent. (1) asserted the existence of sets unconditionally. (5), on the other hand, is completely conditional; first we have to be given the set z, then we can assert the existence of the subset y.

It should be clear that we cannot pass from (5) to a contradiction like (4). Using again the formula '$-(x \in x)$' as an instance of (5) we have:

(6) $$(\exists y)(\forall x)[x \in y \leftrightarrow x \in z \;\&\; -(x \in x)],$$

and again taking $x = y$, we infer:

(7) $$y \in y \leftrightarrow y \in z \;\&\; -(y \in y),$$

which is not contradictory. To make the meaning of (7) a little clearer let z be the set A whose only two members are the set consisting of the number 1 and the set consisting of the number 2, that is,

(8) $$A = \{\{1\}, \{2\}\}.$$

(In (8) we have informally introduced a familiar notation for describing sets: we describe a set by writing down names or descriptions of its members, separated by commas, and enclosing the whole in braces. In the next chapter this notation is formally defined.) Considering now the set A and Russell's formula '$-(x \in x)$', we have from the axiom schema of separation:

(9) $$(\exists y)[y \in y \leftrightarrow y \in A \;\&\; -(y \in y)].$$

The truth of (9) is seen by choosing A itself as an appropriate y, for A is not a member of itself. Thus the left-hand side is false, and the right-hand side is also false, since '$A \in A \;\&\; -(A \in A)$' is contradictory.

Both the axiom schema of abstraction and the axiom schema of separation have been stated as though it were perfectly clear exactly what formulas may be substituted for '$\varphi(x)$'. In the next chapter a precise syntactical definition of *formula* is considered. What has been important historically is that it is by means of an exact definition of the formulas of a theory (here set theory) that application of an axiom schema like that of separation may be made precise. Zermelo [1908] originally formulated the axiom schema of separation in terms of questions or statements which have the property of being *definite*. Roughly speaking, he said that a statement is definite if it can be decided in a non-arbitrary manner whether or not any object satisfies the statement.* His formulation of the axiom

---

*The decision does not have to be by some effective or finite procedure. For further elaboration of this point, see Zermelo [1929].

schema is then (slightly paraphrased): If a statement $\varphi(x)$ is definite for all elements of a set M, then there is always a subset $M\varphi$ of M which contains exactly those elements x of M for which $\varphi(x)$ is true.

The first real clarification of this notion of definiteness was given by Skolem [1922], who characterizes definite statements as just those which satisfy his exact definition of formula. For some further discussion see Zermelo [1929] and Skolem [1930].*

It is not feasible to enter into the details of these papers by Zermelo and Skolem, but some readers, not interested in clarity for its own sake, may wonder why Zermelo was so interested in the first place in restricting the axiom schema of separation to *definite* statements. The answer to this query is most easily given in the context of discussing further paradoxes which arise in the foundations of mathematics.

Before turning to these paradoxes in the next section a historical remark may be made about the axioms to be considered in the sequel. Essentially they correspond very closely to those in Zermelo [1908]. However, when we come to the theory of transfinite induction and ordinal arithmetic, we shall need to add a stronger axiom schema than that of separation, namely, what is usually called the *axiom schema of replacement*, which is due to Fraenkel [1922a].† For these reasons the system of axiomatic set theory developed in this book is usually called *Zermelo-Fraenkel set theory* in the literature of the subject, although it would seem historically more appropriate to call it Zermelo-Fraenkel-Skolem set theory.

§ 1.4 More Paradoxes.

Because of its simplicity Russell's paradox was introduced to show why the obvious, direct axiomatization of intuitive set theory is inconsistent. Historically, other paradoxes were discovered before Russell's, the first published one apparently being Burali-Forti's paradox [1897] of the greatest ordinal. A full analysis of the ten or twelve paradoxes which have been discussed in the literature would be out of place here.‡ For a good survey the reader is referred to Beth [1950]. Several of the paradoxes are relatively slight variations of each other, so that we shall briefly and informally describe only the more prominent ones.

F. P. Ramsey [1926] seems to be the first person explicitly and clearly to divide the paradoxes into two classes: the logical or mathematical paradoxes, and the linguistic or semantical ones. Roughly speaking, the first class arises from purely mathematical constructions; the second from direct

---

*Zermelo's paper of 1929 was concerned also with making the notion of definiteness more precise; it was written without knowledge of Skolem's paper of 1922, and does not provide as satisfactory a formulation as Skolem's. Some telling criticisms of Zermelo's paper are made by Skolem in his paper of 1930. A less detailed but essentially correct clarification of *definiteness* was given independently by Fraenkel [1922b].

†Essentially the same axiom was proposed independently and at the same time by Skolem [1922].

‡A paradox is often also called an *antinomy* in the literature.

consideration of the language which we use to talk about mathematics and logic.

Russell's paradox belongs to the first class, as does the Burali-Forti paradox, which is discussed further in §5.3. The general idea of the latter goes as follows. In intuitive set theory every well-ordered set has an ordinal number. Moreover, the set of all ordinals is well-ordered, whence the set of all ordinal numbers has an ordinal number, Θ say. But the set of all ordinals up to and including any given ordinal is well-ordered and thus has an ordinal number, which exceeds the given ordinal by one. Consequently, the set of all ordinals including Θ has the ordinal number $\Theta + 1$, which is greater than Θ. Therefore Θ is not the ordinal number of all ordinals.

It might be thought that any device which blocked direct derivation of Russell's paradox would do the same for Burali-Forti's, but this is not the case. For example, J. B. Rosser [1942] derived the latter paradox in the system of Quine [1940] thereby showing its inconsistency, although it is clear that no direct derivation of Russell's paradox is possible in Quine's system.

Another well-known paradox in the first class is Cantor's paradox of the greatest cardinal number, which he found in 1899 and which was first published with his correspondence in 1932. Again operating in intuitive set theory we consider the cardinal number $\mathfrak{n}$ of the set S of all sets. On the one hand it is clear that $\mathfrak{n}$ is the greatest possible cardinal. But we may also consider the set of all subsets of S, and its cardinal $\mathfrak{p}$. By a standard theorem of intuitive set theory $\mathfrak{p}$ must be greater than $\mathfrak{n}$.

To readers unfamiliar with the notions of cardinal and ordinal number freely used in describing these last two paradoxes it may be said that in subsequent chapters these notions will be developed in detail and completely *ab ovo*. Without giving an exact analysis at this point it should be clear in a general way how these paradoxes do not arise when the primary axiom for constructing sets is Zermelo's axiom schema of separation, for then the existence of the set of all sets or the set of all ordinal numbers cannot be established.

The oldest semantical paradox is Epimenides' paradox of the liar. Epimenides the Cretan said, "I am lying." If the statement is true then he is lying and the statement is false. If the statement is false, then he is not lying and the statement is true. Modern versions often read like this. Consider the sentence, "The only sentence on this panel of the blackboard is false." If the sentence is true it must be false, and conversely.

An amusing related puzzle which dates from antiquity is the dilemma of the crocodile. The crocodile has stolen a child and says to the father, "I will return the child if you guess correctly whether or not I will return the child." The father replies, "You will not return the child." What should the crocodile do?

The first published modern semantical paradox seems to be Richard's paradox [1905], which is related to Cantor's proof of the non-denumerability of the set of all real numbers (which proof is given in §6.6).* By an *expression* of the English language let us mean any finite sequence of the twenty-six letters, a comma, a period, and a blank space; that is, an expression is any finite sequence of these twenty-nine symbols. Now order the expressions according to total number of symbols and lexicographically within each total number. Thus we have

$$a$$
$$b$$
$$\vdots$$
$$aa$$
$$ab$$
$$\vdots$$
$$aaa$$
$$\vdots$$

Now strike out those expressions which do not define real numbers; let E be the remaining subsequence. Translating and paraphrasing somewhat, we may use Richard's original formulation to define a certain real number N with respect to E: "The real number whose whole part is zero, and whose n-th decimal is p plus one if the n-th decimal of the real number defined by the nth member of E is p and p is neither eight nor nine, and is simply one if this n-th decimal is eight or nine." By construction N is not a member of E, since it differs from every real number in E in at least one decimal place, but N is defined by a finite expression, whence it is in E. Thus the contradiction.

The third and final semantical paradox to be mentioned here is the Grelling-Nelson paradox [1908] of *heterologicality*. A predicate is called heterological if the sentence ascribing to the predicate the property expressed by the predicate is false. Thus the predicate 'red' is heterological since the sentence, "The predicate 'red' is red" is false. The contradiction arises from asking if the predicate 'heterological' is itself heterological. Clearly, if it is, we infer that it is not; and if it is not that it is.

A detailed discussion of these semantical paradoxes would take us far from set theory proper into the general domain of formal logic. But it is pertinent to see how the inferences leading to them are blocked in Zermelo-Fraenkel set theory. Zermelo specifically introduced his notion of definiteness in the axiom schema of separation to prevent construction of the semantical paradoxes (cf. Zermelo [1908, p. 264]). As was pointed

*For an analysis and exposition of Richard's paradox, see Church [1934].

out in the preceding section, this notion of definiteness is made precise by reducing it to the syntactical notion of formula. With reference to the semantical paradoxes the point of this reduction may be made clearer. Every one of these paradoxes arises from having available in the language expressions for referring to other expressions in the language. Any language with such unlimited means of expression is perforce inconsistent.* Consequently it is important to distinguish between the object language — here the language in which we talk about sets — and the metalanguage, that is, the language in which we talk about the object language. Although set theory is not developed in this book in a fully formalized manner, at the beginning of the next chapter we consider an exact definition of *formula* for the object language we use. Our metalanguage is a certain vaguely defined fragment of ordinary English augmented by certain symbols familiar from intuitive mathematics. It will be obvious that our object language is not rich enough to provide any direct means of expressing the semantical paradoxes. In other words, we avoid these paradoxes by severely restricting the richness of our language. It should be noted that when a formalized language is used it is intuitively clear there is little prospect of deriving one of the semantical paradoxes in this language; the intuitive status of the mathematical or logical paradoxes is usually not as immediately clear.

These semantical matters will not be pursued in the sequel. They have been discussed in a somewhat cursory fashion here to clarify the need of what amounts to a semantical restriction on the axiom schema of separation. Furthermore, it should be mentioned that a complete and exact formalization of the object language is necessary for proving metamathematical facts about Zermelo-Fraenkel set theory. By 'metamathematical facts' we mean facts about the object language. An example of an important metamathematical fact is that the axiom *schema* of separation cannot be replaced by a finite number of axioms in the object language.

Finally, it may be emphasized that Zermelo-Fraenkel set theory provides but one of several possible approaches to the foundations of mathematics. There is one alternative so intimately tied to Zermelo-Fraenkel set theory that it should be mentioned here, namely, von Neumann-

---

*In Tarski [1956, p. 402] this is put most succinctly: "The main source of the difficulties met with seems to lie in the following: it has not always been kept in mind that the semantical concepts have a relative character, that they must always be related to a particular language. People have not been aware that the language *about which* we speak need by no means coincide with the language *in which* we speak. They have carried out the semantics of a language in that language itself and, generally speaking, they have proceeded as though there was only one language in the world. The analysis of the antinomies mentioned shows, on the contrary, that the semantical concepts simply have no place in the language to which they relate, that the language which contains its own semantics, and within which the usual logical laws hold, must inevitably be inconsistent."

Bernays-Gödel set theory.* There are two essential differences. The latter theory may be finitely axiomatized. No axiom schema of construction like that of separation is required, but instead a finite number of specific set and class constructions suffices. And in what we shall call for brevity von Neumann set theory, there is a technical distinction between classes and sets. Every set is a class, but not conversely. Those classes which are not sets are called *proper classes*, and their distinguishing characteristic is that they are not members of any other class. The class of all ordinal numbers and the class of all sets both exist, but both are proper classes. Thus the Burali-Forti and Cantor paradoxes cannot be constructed, for they require these classes to be members of other classes. Similar remarks apply to Russell's paradox. In subsequent chapters informal indications are often given to indicate the slight variations in theorems, definitions or proofs required in von Neumann set theory. Zermelo and von Neumann set theory are so closely connected that anyone familiar with the one will soon find himself at home in the other.

§ 1.5 Preview of Axioms. Because the axioms for Zermelo-Fraenkel set theory which we shall use are introduced individually in various sections of succeeding chapters, a cursory preview will perhaps help provide a general sense of development. The remarks at this point are superficial for we shall introduce the axioms only by name. In the next chapter we consider the following seven axioms, the bulk of those needed:

> Axiom of extensionality
> Axiom schema of separation
> Union axiom
> Pairing axiom
> Axiom of regularity
> Sum axiom
> Power set axiom

Toward the end of Chapter 2 we shall show that the union axiom is redundant; that is, it may be derived from the other six. It is used at the beginning of the chapter in order to make initial developments simple.

---

*Originally formulated by von Neumann in a series of papers [1925], [1928a], [1929]. His formulation differs considerably from Zermelo set theory because the notion of function is taken as fundamental rather than that of class or set. In a series of papers in the *Journal of Symbolic Logic* Bernays modified the von Neumann approach in order to remain nearer to the original Zermelo system (see References for the list of papers). Bernays introduced two membership relations: one between sets, and one between sets and classes. In Gödel [1940] the theory is still further simplified; his primitive notions are those of set, class, and membership (although membership alone is sufficient). A paper by R. M. Robinson [1937] provides a simplified system close to von Neumann's original one.

In Chapter 3, which is concerned with relations and functions, no new axioms are introduced. In Chapter 4, a special axiom for cardinal numbers is presented and used mainly in the context of that chapter. This special axiom is not a part of classical Zermelo-Fraenkel set theory, but it enormously facilitates the construction of intuitive cardinal number theory within our axiomatic framework.

The axiom of infinity is introduced in Chapter 5 in order to make it possible to prove that the set of all natural numbers exists. The natural numbers themselves can be constructed without this axiom. Chapter 6 is concerned with construction of the real numbers, and no further axioms are needed for this work.

In Chapter 7 the axiom schema of replacement is brought in as necessary for the development of ordinal arithmetic and transfinite induction. It is also shown that the pairing axiom may be derived from this schema and the power set axiom. Chapter 8, the final chapter, is mainly devoted to the axiom of choice.

CHAPTER 2

GENERAL DEVELOPMENTS

§ **2.1 Preliminaries: Formulas and Definitions.** Let us now (a) explicitly define the notion of formula required in the axiom schema of separation (and later in the axiom schema of replacement) and (b) state the approach to definitions we adopt in introducing the many defined symbols needed.

In Chapter 1, the important distinction was made between the object language, that is, the language in which we talk about sets, and the metalanguage, that is, the language in which we discuss the object language itself. We use the metalanguage, which for us is ordinary English augmented by a certain amount of intuitive mathematical language, to describe exactly the object language. It may be helpful to look upon this description as analogous to an exact characterization of a game like chess or bridge. But this analogy is not to be carried too far, for most of the expressions of our object language have a definite meaning in terms of intuitive mathematical ideas, which the positions or moves in a game like chess do not.

We begin with a fivefold classification of the symbols of the object language into constants, variables, sentential connectives, quantifiers or operators, and punctuation or grouping symbols.* The two primitive constants of the language are the membership relation symbol '$\in$', introduced informally in Chapter 1, and the constant '0', which denotes the empty set. In addition, we take from logic the predicate constant '$=$', which is the identity symbol. The general variables ranging over all objects are the letters 'x', 'y', 'z', . . . with and without subscripts or superscripts. The sentential connectives are the five mentioned in §1.2 — $-$, &, $\vee$, $\rightarrow$, $\leftrightarrow$; the three quantifiers or logical operators we use — $\forall$, $\exists$, E! — were also mentioned in §1.2. Finally, left- and right-hand parentheses are our only punctuation symbols.

*This classification originates with von Neumann [1927]. For detailed discussion of these matters see the first chapter of Church [1956].

Expressions of the object language are finite sequences of the five classes of symbols of the language. Certain of these expressions, simply because of their structure, are called *primitive formulas* of the object language. We now define such formulas so that merely by looking at the form of an expression we can automatically decide in a finite number of steps whether or not it is a primitive formula. Although this definition is purely syntactical or structural, it is just the expressions satisfying it which have a clear intuitive meaning. An expression like $'(\rightarrow \in x'$ is not a primitive formula and has no intuitive meaning.

We first define primitive *atomic* formulas.

A primitive atomic formula is an expression of the form $(\mathbf{v} \in \mathbf{w})$, *or of the form* $(\mathbf{v} = \mathbf{w})$, *where* $\mathbf{v}$ *and* $\mathbf{w}$ *are either general variables or the constant* '0'.*

Thus $'(x \in y)'$ and $'(z = 0)'$ are primitive atomic formulas.

We may now give what is usually called a recursive definition of *primitive formulas:*

(a)   *Every primitive atomic formula is a primitive formula;*

(b)   *If* $\mathbf{P}$ *is a primitive formula, then* $-\mathbf{P}$ *is a primitive formula;*

(c)   *If* $\mathbf{P}$ *and* $\mathbf{Q}$ *are primitive formulas, then* $(\mathbf{P} \mathbin{\&} \mathbf{Q})$, $(\mathbf{P} \vee \mathbf{Q})$, $(\mathbf{P} \rightarrow \mathbf{Q})$, *and* $(\mathbf{P} \leftrightarrow \mathbf{Q})$ *are primitive formulas;*

(d)   *If* $\mathbf{P}$ *is a primitive formula and* $\mathbf{v}$ *is any general variable then* $(\forall\mathbf{v})\mathbf{P}$, $(\exists\mathbf{v})\mathbf{P}$ *and* $(\mathrm{E}!\mathbf{v})\mathbf{P}$ *are primitive formulas;*

(e)   *No expression of the object language is a primitive formula unless its being so follows from rules* (a) – (d).

The following are examples of primitive formulas of the object language which are not atomic: $'(\exists x)(\forall y) - (y \in x)'$, $'((x \in y) \rightarrow (y \in z))'$, $'(\mathrm{E}!\,z)$ $(0 = z)'$. In terms of this definition, an exact formulation of the axiom schema of separation is then:

Any primitive formula of the object language of the form

$$(\exists\mathbf{v})((\exists\mathbf{w}_1)(\mathbf{w}_1 \in \mathbf{v} \vee \mathbf{v} = 0) \mathbin{\&} (\forall\mathbf{w})(\mathbf{w} \in \mathbf{v} \leftrightarrow \mathbf{w} \in \mathbf{u} \mathbin{\&} \varphi))$$

is an axiom, provided the variable $\mathbf{v}$ *is distinct from* $\mathbf{u}$ *and* $\mathbf{w}_1$ *and is not free in the primitive formula* φ.

*In this definition, as elsewhere, we use the boldface letters '$\mathbf{u}$', '$\mathbf{v}$', '$\mathbf{w}$', '$\mathbf{u}_1$', '$\mathbf{v}_1$', '$\mathbf{w}_1$', . . . as metamathematical variables which take as values variables 'x', 'y', 'z', . . . or the constant '0' of the object language. And we use boldface letters, '$\mathbf{P}$', '$\mathbf{Q}$', . . . , as well as Greek letters 'φ' and 'Ψ', as metamathematical variables which take as values formulas of the object language. The conventions about use and mention followed here, which are probably obvious, are that (i) the constants '$\in$' and '$=$', the sentential connectives, the quantifier symbols, and the left and right parentheses are used as names of themselves, and (ii) juxtaposition of names of expressions denotes a binary operation on expressions which yields new expressions (for example, '$x{\in}y$' & '$y{\in}z$' = '$x{\in}y$ & $y{\in}z$'). For a more detailed discussion of these conventions, see Chapter 6 of Suppes [1957].

The first clause of the axiom guarantees that an arbitrary individual cannot play the role of the empty set 0. Justifying reasons for the restrictions on the variable v are given in the next section.

In principle all of the axioms and theorems of set theory that we state in the following pages can be written as primitive formulas of the object language — indeed, our official object language shall consist of these primitive formulas. For working purposes it will be useful and convenient to introduce by definition considerable additional notation. We shall in practice apply the axiom schema of separation to formulas which are not written solely in primitive notation; but since at any point in our development only a finite number of definitions will have preceded, such a formula can be replaced by a primitive formula by a finite number of substitutions.

Regarding definitions, then, our viewpoint is that they are informally admitted if clear recipes are given for eliminating new symbols from any context. We thus require that a formula of the object language which introduces a new symbol must satisfy the following:

> CRITERION OF ELIMINABILITY. *A formula* P *introducing a new symbol satisfies the criterion of eliminability if and only if: whenever* Q_1 *is a formula in which the new symbol occurs, then there is a primitive formula* Q_2 *such that* P → (Q_1 ↔ Q_2) *is derivable from the axioms.*

Notice that we have stated this criterion without giving an exact definition of *formula* (as opposed to *primitive formula*). Such a definition is straightforward if we list all the defined symbols introduced in this book, and then proceed in terms of this list as we did before with the primitive notation. This tedious task we shall not perform, but we do want to mention a second criterion we expect our definitions to satisfy, namely, our definitions must not be creative.

> CRITERION OF NON-CREATIVITY. *A formula* P *introducing a new symbol satisfies the criterion of non-creativity if and only if: there is no primitive formula* Q *such that* P → Q *is derivable from the axioms but* Q *is not.*

In other words, a definition should not function as a creative axiom permitting derivation of some previously unprovable formula in which only primitive notation occurs.

The classical problem of the theory of definition for any exactly stated mathematical theory is to provide rules of definition whose satisfaction entails satisfaction of the two criteria just stated. We may restrict ourselves here to rules for defining operation symbols. Slight modifications yield appropriate rules for defining relation symbols and individual constants.* In these rules we refer to *preceding definitions*, which implies that the definitions are given in a fixed sequence and not simultaneously;

---

*Individual constants may in fact be treated as operation symbols of degree zero.

this approach permits use of defined symbols in the definitions of new symbols.*

Proper definitions of operation symbols may be either equivalences or identities. We begin with the former.

> *An equivalence* P *introducing a new n-place operation symbol* O *is a proper definition if and only if* P *is of the form*
>
> $$O(v_1, \ldots, v_n) = w \leftrightarrow Q$$
>
> *and the following restrictions are satisfied:* (i) $v_1, \ldots, v_n, w$ *are distinct variables,* (ii) Q *has no free variables other than* $v_1, \ldots, v_n, w$, (iii) Q *is a formula in which the only non-logical constants are the primitive or previously defined symbols of set theory, and* (iv) *the formula* (E! w)Q *is derivable from the axioms and preceding definitions.*

Regarding the phrase 'non-logical constants' in (iii), the only logical constants are those introduced in §1.2; all other constants are non-logical. Justification of the various restrictions is easily given. Here we shall emphasize only the importance of (iv). Consider the following definition in elementary arithmetic of the pseudo-operation ⋆.

(1) $$x \star y = z \leftrightarrow x < z \,\&\, y < z.$$

Clearly it is false that

$$(E!z)\ (x < z \,\&\, y < z).$$

Thus (1) violates (iv), and we want to show that violation of it will lead to a contradiction. Now since $1 < 3$, $2 < 3$, $1 < 4$, and $2 < 4$, we immediately infer from (1):

$$1 \star 2 = 3$$

and

$$1 \star 2 = 4$$

whence

$$4 = 3,$$

which is absurd. In ordinary mathematical language the point of (iv) is to require that performing an operation shall always yield a unique object.

For a definition which is an identity we have the following rule.

> *An identity* P *introducing a new n-place operation symbol* O *is a proper definition if and only if* P *is of the form*
>
> $$O(v_1, \ldots, v_n) = t$$

---

*The rules to be given and related matters are discussed in more detail in Chapter 8 of Suppes [1957].

and the following restrictions are satisfied: (i) $v_1, \ldots, v_n$ *are distinct variables,* (ii) *the term* t *has no free variables other than* $v_1, \ldots, v_n$, *and* (iii) *the only non-logical constants in the term* t *are primitive symbols and previously defined symbols of set theory.*

An example of a definition by means of an identity in arithmetic is the definition of subtraction in terms of addition and the negative operation.

$$x - y = x + (-y).$$

It is straightforward to prove that definitions satisfying either of these rules just given, or the analogous ones for relation symbols and individual constants, satisfy the criteria of eliminability and non-creativity.

Unfortunately many of the definitions common in mathematics and many of the definitions to be introduced in the sequel do not satisfy the criterion of eliminability; and thus most of the definitions of operation symbols do not satisfy one of the two rules introduced. The reason for this failure may be simply stated: the definitions are often conditional in form. A typical instance of a conditional definition in arithmetic is provided by a definition of division, for which the problem of division by zero arises.

(1) $$y \neq 0 \rightarrow (x/y = z \leftrightarrow x = y \cdot z).$$

Using (1) as the definition of the operation symbol for division, we cannot eliminate the symbol from contexts like:

$$1/0 \neq 2.$$

On the other hand, we can use (1) to eliminate division in all "interesting" cases, that is, all those which satisfy the hypothesis of (1). Moreover, it is not difficult to modify the two rules given in such a way that conditional definitions satisfying them satisfy the criterion of non-creativity. In fact, the appropriate modifications of the rule for equivalences which define operation symbols are embodied in the following.

An implication P *introducing a new operation symbol* O *is a conditional definition if and only if* P *is of the form*

$$Q \rightarrow [O(v_1, \ldots, v_n) = w \leftrightarrow R]$$

and the following restrictions are satisfied: (i) *the variable* w *is not free in* Q, (ii) *the variables* $v_1, \ldots, v_n$, w *are distinct,* (iii) R *has no free variables other than* $v_1, \ldots, v_n$, w, (iv) Q *and* R *are formulas in which the only non-logical constants are the primitive symbols and previously defined symbols of set theory, and* (v) *the formula* $Q \rightarrow (E!w)R$ *is derivable from the axioms and preceding definitions.*

To convert a conditional definition of an operation symbol into a proper definition satisfying the criterion of eliminability is a routine matter once an object is chosen which can be the result of performing the operation when the hypothesis of the conditional definition is not satisfied. In arithmetic the natural choice is the number zero. Having agreed upon this, we may replace the conditional definition (1) of division by:

$$x/y = z \leftrightarrow (y \neq 0 \rightarrow x = y \cdot z) \ \& \ (y = 0 \rightarrow z = 0).$$

The corresponding natural choice in set theory is the empty set. Thus any conditional definition satisfying the rule stated above may be converted into a proper definition by writing it as:

$$(2) \qquad \mathsf{O}(\mathsf{v}_1, \ldots, \mathsf{v}_n) = \mathsf{w} \leftrightarrow (\mathsf{Q} \rightarrow \mathsf{R}) \ \& \ (-\mathsf{Q} \rightarrow \mathsf{w} = 0).$$

In the sequel we shall continually use conditional definitions, but with the understanding that the notation they introduce may always be eliminated in favor of primitive notation by recasting them as proper definitions of the form indicated by (2).

From a logical standpoint definitions of the kind we have been describing are non-creative axioms stated in the object language. They are properly classified as axioms because they function as additional premises in the derivation of theorems, but their non-creative character insures that they do not actually strengthen set theory as formulated in the basic creative axioms. We shall occasionally introduce definition schemata which, like the axiom schema of separation, should properly be formulated in the metalanguage. Such definition schemata will mainly occur in connection with the introduction of new methods of binding variables. We shall also augment our primitive notation by the introduction of several kinds of variables: set variables ranging over sets but not individuals ('A', 'B', 'C', . . .), cardinal variables ranging over cardinal numbers ('$\mathfrak{m}$', '$\mathfrak{n}$', '$\mathfrak{p}$', . . .), ordinal variables ranging over ordinal numbers ('α', 'β', 'γ', . . .), integer variables ranging over non-negative integers ('m', 'n', 'p', . . .), rational variables ranging over non-negative rational numbers ('M', 'N', 'P', . . .). We shall not explicitly consider rules for definition schemata which introduce new variables or new methods of binding variables, but it will be clear for the small number of such schemata actually used how it may be shown that they satisfy the two criteria of eliminability and non-creativity, or how they may be slightly modified to obtain this satisfaction.

§ 2.2 **Axioms of Extensionality and Separation.** We begin with the definition of the notion of a set. The content of the definition conforms to intuitive ideas: a set is something which has members or is the empty set.

DEFINITION 1.   *y is a set* $\leftrightarrow (\exists x) \ (x \in y \lor y = 0).$

As might be expected, the notion of being a set is needed every step of the way. Most of the definitions introduced, for instance, will be conditional ones intended intuitively to apply only to sets. In order to save the trouble of continually writing out the predicate 'is a set', we shall adopt the following convention regarding variables. Capital italic letters '*A*', '*B*', '*C*', etc. shall be used only for sets. Lower case italic letters '*x*', '*y*', '*z*', etc., on the other hand, may take as their values either sets or individuals. (We call the latter symbols *general variables*.) With this convention clearly in mind we may without any confusion ordinarily omit the predicate 'is a set'.

Translation of sentences involving set variables into the basic notation of general variables is straightforward. The three rules needed are quite simple; rather than give a formal statement of them, we may illustrate their use by three examples, one dealing with a universal quantifier, one with an existential quantifier, and one with the 'exactly one' quantifier. The sentence:

$$(\forall A)(\exists x) - (x \in A)$$

is translated:

$$(\forall y)(y \text{ is a set} \rightarrow (\exists x) - (x \in y)),$$

and the sentence:

$$(\forall x)(\exists A) - (x \in A)$$

is translated:

$$(\forall x)(\exists y)(y \text{ is a set} \ \& \ -(x \in y)).$$

The sentence:

$$(E!A)(\forall x)(x \in A)$$

is translated:

$$(E!y)(y \text{ is a set} \ \& \ (\forall x)(x \in y)).$$

It is to be noted that the individual constant denoting the empty set is used in the definiens of Definition 1. It is customary in many axiomatic developments of set theory to define the symbol for the empty set rather than to begin with it as a primitive symbol. But this is not possible here, for the axioms are framed so as to permit individuals within the domain of discourse. In this respect we follow Zermelo's original 1908 formulation of set theory. However, our axioms do not actually postulate the existence of any individuals, and they are thus consistent with the view that there are only sets in the domain of discourse. A standard definition of the empty set is:

$$0 = x \leftrightarrow (\forall y)(y \notin x),$$

but it is easy to prove on the basis of this definition that any individual is identical with the empty set, and this effectively excludes individuals.

The two axioms we consider in this section are the *axiom of extensionality:*

$$(\forall x)(x \in A \leftrightarrow x \in B) \rightarrow A = B$$

and the *axiom schema of separation:*

$$(\exists B)(\forall x)(x \in B \leftrightarrow x \in A \,\&\, \varphi(x)).$$

It is understood in the axiom schema of separation that the variable 'B' is not free in $\varphi(x)$. An exact metamathematical formulation of this schema was given in the preceding section.* For intuitive working purposes we shall hold to the form just now used, which is a mixture of the object language and metalanguage; but the reader should keep clearly in mind that this is an axiom schema, not a single axiom. The restriction that 'B' not be free in $\varphi(x)$ is essential, for without it we can derive a contradiction whenever A is a non-empty set. To see this, let $\varphi(x)$ be '$-(x \in B)$' and let A be the set consisting of the empty set (the existence of A follows from other axioms in this chapter). Then we have:

$$(\exists B)(0 \in B \leftrightarrow 0 \in A \,\&\, -(0 \in B)),$$

which, since $0 \in A$, implies:

$$(\exists B)(0 \in B \leftrightarrow -(0 \in B)),$$

an absurdity. On the other hand, it is legitimate and sometimes necessary for $\varphi(x)$ to contain free variables other than simply 'x'; only 'B' is excluded.

We turn now to systematic developments. We first define the standard notation '$\not\in$' for something not being a member of something else.

Definition 2. $x \not\in y \leftrightarrow -(x \in y)$.

Correspondingly, we use from logic the notation '$x \neq x$' for '$-(x = x)$'.

As the first theorem, we have:

Theorem 1. $x \not\in 0$.

proof. Taking $\varphi(x)$ as '$x \neq x$' we have from the axiom schema of separation:

$$(1) \qquad (\exists A)(\forall x)(x \in A \leftrightarrow x \in 0 \,\&\, x \neq x).$$

Suppose now that some x is in A; then by (1), $x \neq x$, which is absurd. Hence we conclude:

$$(2) \qquad (\forall x)(x \not\in A),$$

---

*It is worth noting that strictly speaking the formulation with set variables is weaker than the metamathematical one with only one kind of variable, for the former reads:

$$x \text{ is a set} \rightarrow (\exists y)(y \text{ is a set} \,\&\, (\forall z)(z \in y \leftrightarrow z \in x \,\&\, \varphi(z))),$$

whereas the earlier metamathematical version does not have this conditional form. On the other hand, it is of no real interest to have the axiom apply when x is an individual.

and therefore by Definition 1

(3) $A = 0$.

The theorem follows from (2) and (3).   Q.E.D.

We next prove a simple theorem concerning the uniqueness of the empty set.

THEOREM 2. $(\forall x)(x \notin A) \leftrightarrow A = 0$.

PROOF.   If $A = 0$, then by Theorem 1, $x \notin A$.   On the other hand, if for every x, $x \notin A$, then there is no element in the set A and by Definition 1, $A = 0$.   Q.E.D.

The remainder of this section is concerned with the notions of inclusion and proper inclusion of sets.   If A and B are sets such that every member of A is a member of B, then we say that A *is included in* B, or that A *is a subset of* B, which we symbolize: $A \subseteq B$.   Thus we may write:

The set of Irishmen is included in the set of men

or:

The set of Irishmen is a subset of the set of men

or simply:

The set of Irishmen $\subseteq$ the set of men.

Formally, we have:

DEFINITION 3. $A \subseteq B \leftrightarrow (\forall x)(x \in A \rightarrow x \in B)$.

From a formal standpoint Definition 3 is a conditional definition of a two-place relation symbol.   The fact that it is a conditional definition is masked by the use of capital letters in the manner agreed upon.   This same remark applies to all the definitions introduced which use set variables.

THEOREM 3. $A \subseteq A$.

PROOF.   Since it is a truth of logic that

$$(\forall x)(x \in A \rightarrow x \in A),$$

it follows immediately from Definition 3 that

$$A \subseteq A. \qquad\qquad \text{Q.E.D.}$$

Theorem 3 asserts simply that inclusion is reflexive; the next theorem asserts that it has the property of antisymmetry, as it is usually called.

THEOREM 4. $A \subseteq B \ \& \ B \subseteq A \rightarrow A = B$.

PROOF.   If $A \subseteq B$ and $B \subseteq A$, then it follows from Definition 3 that

$$(\forall x)(x \in A \leftrightarrow x \in B).$$

Hence by the axiom of extensionality, $A = B$.   Q.E.D.

THEOREM 5.　$A \subseteq 0 \rightarrow A = 0$.

PROOF.　By virtue of Definition 3 and the hypothesis of the theorem, if $x \in A$ then $x \in 0$. But by Theorem 1 $x \notin 0$. Hence for every x, $x \notin A$, and thus by Theorem 2, $A = 0$.　　　　　　　　　　　　Q.E.D.

Transitivity of inclusion is asserted in the following theorem.

THEOREM 6.　$A \subseteq B \,\&\, B \subseteq C \rightarrow A \subseteq C$.

PROOF.　Consider an arbitrary element x. Since $A \subseteq B$ if $x \in A$ then $x \in B$, but $B \subseteq C$; hence if $x \in B$ then $x \in C$. Thus by the transitivity of implication if $x \in A$ then $x \in C$. Q.E.D.*

A typical procedure used in informal proofs is exemplified here. We need to prove something about all elements x. To do this it suffices to give the argument for an arbitrary x. The use of the phrase 'arbitrary element' corresponds in the language of logic to introducing a free variable in a premise (in this case the premise: $x \in A$).

We now define *proper inclusion*.

DEFINITION 4.　$A \subset B \leftrightarrow A \subseteq B \,\&\, A \neq B$.

Thus using informally the braces notation not yet formally defined, we have:

$$\{1,2\} \subset \{1,2,3\}$$

but it is not the case that

$$\{1,2\} \subset \{1,2\}.$$

(Here we describe a set by writing down names of its members separated by commas and enclose the result with curly braces.) The next four theorems assert expected properties of proper inclusion; the proofs are left as exercises.

THEOREM 7.　$-(A \subset A)$

THEOREM 8.　$A \subset B \rightarrow -(B \subset A)$

THEOREM 9.　$A \subset B \,\&\, B \subset C \rightarrow A \subset C$.

THEOREM 10.　$A \subset B \rightarrow A \subseteq B$.

EXERCISES

1.　Prove Theorems 7-10.

2.　Formulate Definition 4 as a conditional definition without using set variables 'A', 'B', etc.

---

*"Transitivity of implication" just means that from $\mathbf{P} \rightarrow \mathbf{Q}$ and $\mathbf{Q} \rightarrow \mathbf{R}$ we may infer $\mathbf{P} \rightarrow \mathbf{R}$.

3. Formulate Theorems 2 and 3 without using any set variables.

4. Along the lines of Definition 1 give a formal definition of the notion of an individual.

§ 2.3 **Intersection, Union, and Difference of Sets.**   The elementary properties of the three most basic binary operations on sets are the concern of this section.   Using again informally the braces notation for sets, some simple examples illustrating these operations may be given.   If A and B are sets, then by the *intersection* of A and B (in symbols: $A \cap B$) we mean the set of all things which belong both to A and B.   Thus

$$\{1,2\} \cap \{2,3\} = \{2\}$$

and

$$\{1\} \cap \{2\} = 0.$$

By the *union* of A and B (in symbols: $A \cup B$) we mean the set of all things which belong to at least one of the sets A and B.   For instance,

$$\{1,2\} \cup \{2,3\} = \{1,2,3\}$$

and

$$\{1\} \cup \{2\} = \{1,2\}.$$

By the *difference* of A and B (in symbols: $A \sim B$) we mean the set of all things which belong to A but not to B.   Thus

$$\{1,2\} \sim \{2,3\} = \{1\}$$

and

$$\{1\} \sim \{2\} = \{1\}.$$

The basic theorems asserting existence of the sets which are the intersection and difference of two sets may be proved by use of the axiom schema of separation.   The same thing cannot be said for the union of two sets, and so at this point we introduce the *union axiom*, which we will show later to be redundant in terms of the full set of axioms introduced in this Chapter:*

$$(\exists C)(\forall x)(x \in C \leftrightarrow x \in A \lor x \in B).$$

We do not use this axiom immediately, but first develop the properties of the intersection operation.

THEOREM 11. $(E!C)(\forall x)(x \in C \leftrightarrow x \in A \,\&\, x \in B).$

---

*The sum axiom introduced below in §2.6 is sometimes called the union axiom.   The present axiom, which is redundant, must not be confused with it.

PROOF.   The following is an instance of the axiom schema of separation:

$$(\exists C)(\forall x)(x \in C \leftrightarrow x \in A \ \& \ x \in B).$$

We now need to show that C is unique. Suppose there were a second set C' such that for every x

$$x \in C' \leftrightarrow x \in A \ \& \ x \in B;$$

then for every x

$$x \in C' \leftrightarrow x \in C$$

and by virtue of the axiom of extensionality

$$C' = C. \hspace{3cm} \text{Q.E.D.}$$

The theorem just proved formally justifies the definition of intersection.

DEFINITION 5. $A \cap B = y \leftrightarrow (\forall x)(x \in y \leftrightarrow x \in A \ \& \ x \in B) \ \& \ y \text{ is}$ *a set.*

Admittedly the natural tendency is to write in place of Definition 5, the formula:

(1) $\hspace{2cm} A \cap B = C \leftrightarrow (\forall x)(x \in C \leftrightarrow x \in A \ \& \ x \in B),$

but (1) does not translate back into general variables in a satisfactory manner; for it becomes:

(2) x,y,z are sets $\rightarrow (x \cap y = z \leftrightarrow (\forall w) \ (w \in z \leftrightarrow w \in x \ \& \ w \in y)),$

and the reasons for preventing the free occurrence of 'z' in the hypothesis of (2) are obvious; if it is there we cannot prove, for instance, that $0 \cap 0 \neq z$ for any individual z.   Note that the rule for conditional definitions given in the preceding section prohibited such a free occurrence of 'z'.

If the definitional rules stated at the beginning of this chapter had been more liberal, we could have adopted as the definition of intersection the following theorem, which has the virtue of being much easier to work with in proofs. The argument against liberalizing the rules of definition is that non-creativity of definitions becomes much less obvious and in many cases difficult to prove.

THEOREM 12. $x \in A \cap B \leftrightarrow x \in A \ \& \ x \in B.$

PROOF.   Using the identity:

$$A \cap B = A \cap B$$

and putting '$A \cap B$' for 'y' in Definition 5, we obtain the theorem at once.

$$\text{Q.E.D.}$$

The next two theorems assert the commutativity and associativity of intersection. The proofs are left as exercises.

THEOREM 13. $A \cap B = B \cap A$.

THEOREM 14. $(A \cap B) \cap C = A \cap (B \cap C)$.

A binary operation is idempotent if the operation, when performed on any element with itself, results in just that element. The next theorem asserts the idempotence of intersection.

THEOREM 15. $A \cap A = A$.

PROOF.   By Theorem 12

$$x \in A \cap A \leftrightarrow x \in A \,\&\, x \in A,$$

but

$$x \in A \,\&\, x \in A \leftrightarrow x \in A,$$

and thus by Definition 5

$$A \cap A = A. \hspace{3cm} \text{Q.E.D.}$$

Three intuitively obvious theorems are next stated; only the first is proved.

THEOREM 16. $A \cap 0 = 0$.

PROOF.   By virtue of Theorem 12

$$x \in A \cap 0 \leftrightarrow x \in A \,\&\, x \in 0,$$

but by Theorem 1

$$x \notin 0.$$

Hence

$$x \notin A \cap 0,$$

and since the argument holds for every x, by Theorem 2

$$A \cap 0 = 0. \hspace{3cm} \text{Q.E.D.}$$

THEOREM 17. $A \cap B \subseteq A$.

THEOREM 18. $A \subseteq B \leftrightarrow A \cap B = A$.

We now turn to the theorem justifying the operation of union of sets. The proof of this theorem involves the first use of the union axiom.

THEOREM 19. $(\text{E}!C)(\forall x)(x \in C \leftrightarrow x \in A \lor x \in B)$.

PROOF.   Similar to the proof of Theorem 11, but using the union axiom in place of the axiom schema of separation.

DEFINITION 6. $A \cup B = y \leftrightarrow (\forall x)(x \in y \leftrightarrow x \in A \lor x \in B) \,\&\, y$ *is a set.*

For working purposes we immediately need a theorem for the operation of union analogous to Theorem 12.

THEOREM 20. $x \in A \cup B \leftrightarrow x \in A \vee x \in B.$

PROOF.   Similar to the proof of Theorem 12.

Since many of the proofs of theorems about union of sets parallel those about intersection, we may often dismiss them with reference to the corresponding theorem about intersection, as we have done in the proofs of Theorems 19 and 20.

The next three theorems assert the commutativity, associativity, and idempotence of union.

THEOREM 21. $A \cup B = B \cup A.$

THEOREM 22. $(A \cup B) \cup C = A \cup (B \cup C).$

THEOREM 23. $A \cup A = A.$

Further facts are asserted in the next four theorems.

THEOREM 24. $A \cup 0 = A.$

THEOREM 25. $A \subseteq A \cup B.$

THEOREM 26. $A \subseteq B \leftrightarrow A \cup B = B.$

THEOREM 27. $A \subseteq C \ \& \ B \subseteq C \to A \cup B \subseteq C.$

We now state two fundamental distributive laws for intersection and union, and prove the first one.

THEOREM 28. $(A \cup B) \cap C = (A \cap C) \cup (B \cap C).$

PROOF.   Let x be an arbitrary element. By virtue of Theorem 12

$$x \in (A \cup B) \cap C \leftrightarrow x \in A \cup B \ \& \ x \in C,$$

and by Theorem 20

$$x \in A \cup B \ \& \ x \in C \leftrightarrow (x \in A \vee x \in B) \ \& \ x \in C,$$

and by the distributive laws of sentential logic*

$$(x \in A \vee x \in B) \ \& \ x \in C \leftrightarrow (x \in A \ \& \ x \in C) \vee (x \in B \ \& \ x \in C).$$

Using again Theorem 12

$$(x \in A \ \& \ x \in C) \vee (x \in B \ \& \ x \in C) \leftrightarrow x \in A \cap C \vee x \in B \cap C,$$

*The law in question is that from (P v Q) & R we may infer (P & R) v (Q & R) and conversely, where P, Q, and R are any formulas.

and now using Theorem 20 again

$$x \in A \cap C \lor x \in B \cap C \leftrightarrow x \in (A \cap C) \cup (B \cap C).$$

Hence from the transitivity of equivalence we infer:

$$x \in (A \cup B) \cap C \leftrightarrow x \in (A \cap C) \cup (B \cap C),$$

and thus by the axiom of extensionality

$$(A \cup B) \cap C = (A \cap C) \cup (B \cap C). \qquad \text{Q.E.D.}$$

In the proof of Theorem 28 a device which is used over and over again has been employed: in order to prove two sets identical, we begin by considering an arbitrary element of one of the sets and show that it belongs to this set if and only if it belongs to the other. Using the axiom of extensionality we immediately obtain the identity of the two sets.

THEOREM 29. $(A \cap B) \cup C = (A \cup C) \cap (B \cup C).$

We next state the justifying theorem and definition of the operation of set difference.

THEOREM 30. $(E!C)(\forall x)(x \in C \leftrightarrow x \in A \,\&\, x \notin B).$

PROOF. Similar to Theorem 11, but here taking $\varphi(x)$ as '$x \notin B$'.

DEFINITION 7. $A \sim B = y \leftrightarrow (\forall x)(x \in y \leftrightarrow x \in A \,\&\, x \notin B) \,\&\, y$ *is a set.*

THEOREM 31. $x \in A \sim B \leftrightarrow x \in A \,\&\, x \notin B.$

PROOF. Similar to the proof of Theorem 12.

The next theorem obviously entails the fact that difference of sets is not idempotent, granted the existence of non-empty sets.

THEOREM 32. $A \sim A = 0.$

PROOF. By Theorem 1

$$x \notin 0.$$

Hence by sentential logic

$$x \in 0 \leftrightarrow x \in A \,\&\, x \notin A,$$

and thus by Definition 7

$$A \sim A = 0. \qquad \text{Q.E.D.}$$

The remaining theorems of this section state facts relating the set operations of intersection, union and difference. The proofs of the theorems are easily given by employing an approach similar to that used for Theorem

28. As with that theorem, the proofs depend upon exploiting formal properties of the sentential connectives analogous to the formal properties asserted in the theorems. The proof of only the first of these theorems is given here.

THEOREM 33. $A \sim (A \cap B) = A \sim B$.

PROOF.   Let x be an arbitrary element. Then

$$x \in A \sim (A \cap B) \leftrightarrow x \in A \,\&\, -(x \in A \cap B) \qquad \text{by Theorem 31}$$

$$\leftrightarrow x \in A \,\&\, -(x \in A \,\&\, x \in B) \qquad \text{by Theorem 12}$$

$$\leftrightarrow x \in A \,\&\, (x \notin A \vee x \notin B) \qquad \text{by sentential logic}$$

$$\leftrightarrow (x \in A \,\&\, x \notin A) \vee (x \in A \,\&\, x \notin B) \qquad \text{by sentential logic}$$

$$\leftrightarrow x \in A \,\&\, x \notin B \qquad \text{by sentential logic.}$$
$$\text{Q.E.D.}$$

A staccato style, which consists of displaying a series of equivalences and is similar to that often used for a chain of identities, has been adopted here. Some readers may find this method of presentation clearer than the more prolix one used in the proof of Theorem 28.

THEOREM 34. $A \cap (A \sim B) = A \sim B$.

THEOREM 35. $(A \sim B) \cup B = A \cup B$.

THEOREM 36. $(A \cup B) \sim B = A \sim B$.

THEOREM 37. $(A \cap B) \sim B = 0$.

THEOREM 38. $(A \sim B) \cap B = 0$.

THEOREM 39. $A \sim (B \cup C) = (A \sim B) \cap (A \sim C)$.

THEOREM 40. $A \sim (B \cap C) = (A \sim B) \cup (A \sim C)$.

In von Neumann set theory the universe V, which is the class of all sets, exists. The complement $\sim A$ of a set A can then be defined as

$$\sim A = V \sim A.$$

But this is not possible in Zermelo-Fraenkel set theory, and it may be of interest to see why not in some detail. Analogous to the justifying theorems for the definitions of the three operations already considered, we would need to prove:

(1) $\qquad\qquad (E!B)(\forall x)(x \in B \leftrightarrow x \notin A)$

and then we would define complementation by:

(2) $\sim A = y \leftrightarrow (\forall x)(x \in y \leftrightarrow x \notin A)$ & y is a set.

Suppose now that it were possible to prove (1). Let $A = 0$, then

(3) $(E!B)(\forall x)(x \in B)$,

that is, B is the universal set to which every object belongs; but with B at hand the axiom schema of separation reduces to the axiom schema of abstraction by taking A as the universal set B, and Russell's paradox may be derived as in §1.3. We conclude that (1) cannot be proved and Definition (2) is impossible in Zermelo-Fraenkel set theory. One aspect of this discussion may be formalized in the useful result that there does not exist a universal set. As just indicated, the proof of this theorem proceeds by the line of argument of Russell's paradox via a *reductio ad absurdum*.

THEOREM 41. $-(\exists A)(\forall x)(x \in A)$.

EXERCISES

1. Prove Theorems 13 and 14.
2. Prove Theorems 17 and 18.
3. Prove Theorems 21, 22, and 23.
4. Prove Theorems 24, 25, 26, and 27.
5. Can you give an example of an operation of ordinary arithmetic which is idempotent?
6. Prove Theorem 29.
7. Prove Theorems 34, 35, and 36.
8. Prove Theorems 37 and 38.
9. Prove Theorems 39 and 40.
10. Give a detailed proof of Theorem 41.
11. Find an identity which will serve as a definition of intersection in terms of difference.
12. We define the operation of *symmetric difference* by the identity:

$$A \div B = (A \sim B) \cup (B \sim A).$$

Prove:

(a) $A \div 0 = A$
(b) $A \div B = B \div A$
(c) $(A \div B) \div C = A \div (B \div C)$
(d) $A \cap (B \div C) = (A \cap B) \div (A \cap C)$
(e) $A \sim B \subseteq A \div B$
(f) $A = B \leftrightarrow A \div B = 0$
(g) $A \div C = B \div C \rightarrow A = B$
(h) Find an identity which will serve as a definition of difference in terms of symmetric difference and intersection.

§ 2.4 Pairing Axiom and Ordered Pairs.

The three axioms considered so far permit us to prove the existence of only one set — the empty set. We now introduce the axiom which asserts that given any two elements, that is, any two sets or individuals, we may form the set consisting of these two elements. This axiom is usually called the *pairing axiom:*

$$(\exists A)(\forall z)(z \in A \leftrightarrow z = x \vee z = y).$$

If we wanted to make the union axiom a more integral part of our system, we could replace the pairing axiom by the weaker axiom that the unit set consisting of any element exists. The pairing axiom follows from considering the union of two unit sets. However, as already remarked at the end of Chapter 1, we want to show toward the end of this chapter that the union axiom may be derived from the pairing axiom and the sum axiom, which has not yet been introduced.

Preparatory to the definition of pair sets we have the usual strengthening of the axiom as expressed in the following theorem.

THEOREM 42. $(E!A)(\forall z)(z \in A \leftrightarrow z = x \vee z = y).$

PROOF.   Similar to that of Theorem 11.

DEFINITION 8. $\{x,y\} = w \leftrightarrow (\forall z)(z \in w \leftrightarrow z = x \vee z = y)$ & *w is a set.*

The customary theorem now follows.

THEOREM 43. $z \in \{x,y\} \leftrightarrow z = x \vee z = y.$

PROOF.   Similar to that of Theorem 12.

A less trivial but also useful theorem on unordered pair sets now follows.

THEOREM 44. $\{x,y\} = \{u,v\} \rightarrow (x = u \ \& \ y = v) \vee (x = v \ \& \ y = u).$

PROOF.   By virtue of Theorem 43

$$u \in \{u,v\},$$

and thus by the hypothesis of the theorem

$$u \in \{x,y\}.$$

Hence, by virtue of Theorem 43 again

(1) $$\qquad u = x \vee u = y.$$

By exactly similar arguments

(2) $$\qquad v = x \vee v = y,$$

(3) $$\qquad x = u \vee x = v,$$

(4) $$\qquad y = u \vee y = v.$$

We may now consider two cases.

Case 1. $x = y$. Then by virtue of (1), $x = u$, and by virtue of (2) $y = v$.

Case 2. $x \neq y$. In view of (1), either $x = u$ or $y = u$. Suppose $x \neq u$. Then $y = u$ and by (3) $x = v$. On the other hand, suppose $y \neq u$. Then $x = u$ and by (4) $y = v$.   Q.E.D.

It is convenient for subsequent use to define unit sets, triplet sets and quadruplet sets. The naturalness of the definitions is at once apparent.

DEFINITION 9. $\{x\} = \{x,x\}$

$$\{x,y,z\} = \{x,y\} \cup \{z\}$$

$$\{x,y,z,w\} = \{x,y\} \cup \{z,w\}.$$

As an immediate corollary of Theorem 44 we have an intuitively obvious theorem about unit sets. The proof is left as an exercise.

THEOREM 45. $\{x\} = \{y\} \to x = y$.

We are now in a position to define ordered pairs in terms of unit sets and unordered pair sets. This definition, which was historically important in reducing the theory of relations to the theory of sets, is due to Kuratowski [1921], but the earliest definition permitting this reduction is to be found in Wiener [1914].

DEFINITION 10. $\langle x,y \rangle = \{\{x\}, \{x,y\}\}$.

Within set theory, as we shall see in the next chapter, relations are defined as sets of ordered pairs. Without something like the present definition at hand it is impossible to develop the theory of relations unless the notion of ordered pairs is taken as primitive. Essentially our only intuition about an ordered pair is that it is an entity representing two objects in a given order. The following theorem establishes that Definition 10 is adequate with respect to this idea; namely, two ordered pairs are identical only when the first member of one is identical with the first member of the other, and similarly for the two second members.

THEOREM 46. $\langle x,y \rangle = \langle u,v \rangle \to x = u \;\&\; y = v$.

PROOF.   By virtue of Definition 10 and the hypothesis of the theorem

$$\{\{x\}, \{x,y\}\} = \{\{u\}, \{u,v\}\},$$

and thus by Theorem 44

(1) $(\{x\} = \{u\} \;\&\; \{x,y\} = \{u,v\}) \vee (\{x\} = \{u,v\} \;\&\; \{x,y\} = \{u\}).$

Suppose the first alternative of (1) holds. Then since

$$\{x\} = \{u\}$$

by Theorem 45

$$x = u,$$

and hence by Theorem 44 and the supposition that $\{x,y\} = \{u,v\}$

$$y = v,$$

which establishes the desired result.

Suppose now that the second alternative of (1) holds. Then since $\{x\} = \{x,x\}$, by Theorem 44

$$x = u \ \& \ x = v,$$

and similarly

$$x = u \ \& \ y = u.$$

Hence

$$x = u \ \& \ y = v. \quad \text{Q.E.D.}$$

Ordered pairs play a prominent role in the section on Cartesian products in this chapter and throughout the next chapter, which is concerned with relations and functions.

EXERCISES

1. Prove Theorem 45.
2. Prove that

$$x = y \rightarrow \langle x,y \rangle = \{\{x\}\}.$$

3. Is it always true that if $\{x,y,z\} = \{x,y,w\}$ then $z = w$?
4. Show, by proving a theorem like Theorem 46, that the following definition of ordered pairs is adequate:

$$\langle x,y \rangle = \{\{x,0\} \ , \ \{y, \{0\}\}\}.$$

§ 2.5 Definition by Abstraction.

In many branches of modern mathematics it is customary to use the notation:

$$\{x \colon \varphi(x)\}$$

to designate the set of all objects having the property φ. For example,

$$\{x \colon x > \sqrt{2}\}$$

is the set of all real numbers greater than $\sqrt{2}$; as another example,

$$\{x \colon 1 < x < 4 \ \& \ x \text{ is an integer}\} = \{2,3\}.$$

It should be clear why use of this notation is called *definition by abstraction*. We begin by considering some property, such as being greater than $\sqrt{2}$, and abstract from this property the *set* of all entities having the property.

Our objective is to give a formal definition of this abstraction operation, but it should be noted that in defining it we are introducing neither a new relation symbol, operation symbol, nor individual constant. What we are introducing is an operator which provides a new method of binding variables. Thus in the expression

$$\{x : x > \sqrt{2}\}$$

the notation

$$\{ - : - \}$$

binds the variable 'x'.

DEFINITION SCHEMA 11.

$$\{x : \varphi(x)\} = y \leftrightarrow [(\forall x)(x \in y \leftrightarrow \varphi(x)) \ \& \ y \ is \ a \ set]$$
$$\lor [y = 0 \ \& -(\exists B)(\forall x)(x \in B \leftrightarrow \varphi(x))].$$

It is immediately clear from the definition that $\{x : \varphi(x)\}$ is a set. The point of the second member of the disjunction of the definiens is to put $\{x : \varphi(x)\}$ equal to the empty set if there is no non-empty set having as members just those entities with property φ. The manner of translating formulas in which set variables are bound by abstraction is straightforward. Thus the schematic formula:

$$\{A : \varphi(A)\}$$

is translated:

$$\{x : x \ is \ a \ set \ \& \ \varphi(x)\}.$$

There are a number of intuitively obvious theorem schemata about the abstraction operation, some of which we now state and prove.

THEOREM SCHEMA 47. $y \in \{x : \varphi(x)\} \rightarrow \varphi(y).$

PROOF.   If $y \in \{x : \varphi(x)\}$, then

$$\{x : \varphi(x)\} \neq 0,$$

and so by Definition Schema 11

$$y \in \{x : \varphi(x)\} \leftrightarrow \varphi(y);$$

we may conclude from the hypothesis of our theorem: $\varphi(y)$.   Q.E.D.

THEOREM 48. $A = \{x : x \in A\}.$

PROOF.   It is a truth of logic that

$$(\forall x)(x \in A \leftrightarrow x \in A).$$

Thus taking $\varphi(x)$ in Definition 11 as '$x \in A$', we obtain the theorem at once.   Q.E.D.

THEOREM 49. $0 = \{x: x \neq x\}$.

PROOF. Suppose there were a y such that

$$y \in \{x: x \neq x\}.$$

Then by Theorem 47

(1) $\qquad\qquad y \neq y,$

which is absurd. Q.E.D.

Similar to Theorem 41, we also have:

THEOREM 50. $0 = \{x: x = x\}$.

We may prove as theorems simple formulas which could be used to define intersection, union, and difference of sets.

THEOREM 51. $A \cap B = \{x: x \in A \ \& \ x \in B\}$.

PROOF. Use Theorem 11 and Definition 11.

THEOREM 52. $A \cup B = \{x: x \in A \vee x \in B\}$.

THEOREM 53. $A \sim B = \{x: x \in A \ \& \ x \notin B\}$.

A point of methodological interest is that if these last three theorems are used as definitions of the three operations, no justifying theorem is needed prior to the definition. (Note that the requirement of such a theorem is not needed in the rule for defining operation symbols by identities in §2.1.). From Definition 11 we know that if the intuitively appropriate set of elements does not exist, then the result of performing the operation is the empty set. However, in order to do any serious work with the operations we need the existence of the intuitively appropriate set, which comes down to saying that when we define operations by abstraction the justifying theorems may come *after* rather than before the definition. This point is illustrated below in §2.7. Definitions which do not need a justifying theorem are often called *axiom-free*.

In the sequel it is convenient to have a more flexible form of definition by abstraction than that provided by Definition 11. In particular we want to be able to put complicated terms before the colon rather than simply single variables. For example, in §2.8 we define the Cartesian product of two sets by:

(1) $\qquad\qquad A \times B = \{\langle x,y\rangle: x \in A \ \& \ y \in B\},$

but on the basis of Definition 11 we need to replace (1) by the more awkward expression:

$$A \times B = \{x: (\exists y)(\exists z)(y \in A \ \& \ z \in B \ \& \ x = \langle y,z\rangle)\}.$$

In the style of Definition 11, we have:

DEFINITION SCHEMA 12.

$$\{\tau(x_1, \ldots ,x_n) : \varphi(x_1, \ldots ,x_n)\} = \{y : (\exists x_1) \ldots (\exists x_n)(y = \tau(x_1, \ldots ,x_n)$$
$$\& \ \varphi(x_1, \ldots ,x_n))\}.$$

It is apparent that Definition Schemata 11 and 12 differ from the other definitions introduced thus far in that both of them are schemata, which should receive a metamathematical formulation. For example, Definition 12 could be given the following form:

> *If* (i) $v_1, \ldots ,v_n,$ w *are any distinct variables,* (ii) $\tau(v_1, \ldots ,v_n)$ *is any term in which no bound variables occur and exactly* $v_1, \ldots ,v_n$ *occur free, and* (iii) w *does not occur in the formula* φ, *then the identity* $\{\tau(v_1, \ldots ,v_n) : \varphi\} = \{w : (\exists v_1) \ldots (\exists v_n) (w = \tau(v_1, \ldots ,v_n) \& \varphi)\}$ *holds.*

Clauses (i)-(iii) make clear the restrictions imposed on Definition 12. Actually it is not necessary to require that τ have no bound variables and in some contexts it might be inconvenient, although not in any cases which arise in this book.

When using either Definition 11 or 12, we shall usually refer to *definition by abstraction* rather than the numbered definition. We close this section with a theorem schema which expresses the important idea that equivalent properties are extensionally identical. The proof is left as an exercise.

THEOREM SCHEMA 54. $(\forall x)(\varphi(x) \leftrightarrow \Psi(x)) \rightarrow \{x : \varphi(x)\} = \{x : \Psi(x)\}.$

EXERCISES

1. In the interests of complete explicitness Definition Schema 11 should have been preceded by the following theorem:

$(E!y)[((\forall x) \ (x \in y \leftrightarrow \varphi(x)) \ \& \ y \text{ is a set}) \lor (y = 0 \ \& \ {-}(\exists B)(\forall x)(x \in B \leftrightarrow \varphi(x)))].$

2. Why can the implication sign in Theorem 47 not be replaced by an equivalence sign $\leftrightarrow$?

3. Prove Theorem 50.

4. Give a detailed proof of Theorem 51.

5. Prove Theorems 52 and 53.

6. Define by abstraction the unordered pair set.

7. Prove Theorem 54.

8. Give a counterexample to the following:

$$(\forall x)(\varphi(x) \rightarrow \Psi(x)) \rightarrow \{x : \varphi(x)\} \subseteq \{x : \Psi(x)\}.$$

9. Theorems 49 and 50 suggest the following principle:

$$(\forall x)(\varphi(x) \leftrightarrow {-}\Psi(x)) \rightarrow \{x : \varphi(x)\} = \{x : \Psi(x)\}.$$

If it is true, prove it. If not, give a counterexample.

§ 2.6   Sum Axiom and Families of Sets.   The sum axiom, which is the basis of the present section, postulates the existence of the union of a family of sets.*   To illustrate the notation, let

$$A = \{\{1,2\}, \{2,3\}, \{4\}, \text{Jane Austen}\}.$$

Then

$$\bigcup A = \{1,2,3,4\}.$$

Here A is a family of sets together with one individual. The *union* or *sum* of A (in symbols: $\bigcup A$) is the set of all things which belong to some member of A. Note that any individual in A is irrelevant to $\bigcup A$. In determining $\bigcup A$ we need only consider those members of A which are non-empty sets.

The formal definition uses the abstraction notation introduced in the preceding section.

DEFINITION 13. $\bigcup A = \{x \colon (\exists B)(x \in B \ \& \ B \in A)\}.$

We know from the characteristics of definition by abstraction that if the appropriate set of elements does not exist then $\bigcup A$ is simply the empty set. However, in order to do any serious work with the union of a family of sets we need the existence of the intuitively appropriate set. To this end we introduce the *sum axiom:*

$$(\exists C)(\forall x)(x \in C \leftrightarrow (\exists B)(x \in B \ \& \ B \in A)).$$

We have at once as a consequence of the axiom and the definition of $\bigcup A$ the desired theorem:

THEOREM 55. $x \in \bigcup A \leftrightarrow (\exists B)(x \in B \ \& \ B \in A).$

The more obvious elementary properties of the $\bigcup$ operation are stated in the next few theorems. Some of the proofs are left as exercises.

THEOREM 56. $\bigcup 0 = 0.$

PROOF.   By Theorem 1

$$-(\exists B)(B \in 0).$$

Hence by Theorem 55 for every x

$$x \notin \bigcup 0. \hspace{4cm} \text{Q.E.D.}$$

THEOREM 57. $\bigcup \{0\} = 0.$

PROOF.   If $B \in \{0\}$ then

$$B = 0$$

---

*The word 'family' is a common synonym for 'set', and we use *family of sets* and *set of sets* interchangeably.

and then

$$x \notin B.$$

Hence by Theorem 55, for every x

$$x \notin \cup \{0\}. \qquad \text{Q.E.D.}$$

If it were postulated that there are no individuals, that is, that every object is a set, then we could prove that if $\cup A = 0$, then either $A = 0$ or $A = \{0\}$. As it is, the sums of many different sets may be empty, in fact the sum of any set whose only members are individuals and the empty set.

THEOREM 58. $\cup \{A\} = A.$

THEOREM 59. $\cup \{A,B\} = A \cup B.$

PROOF. By virtue of the fundamental property of unordered pair sets if $C \in \{A,B\}$ then

$$C = A \vee C = B.$$

Thus by Theorem 55

(1) $\qquad\qquad x \in \cup \{A,B\} \leftrightarrow x \in A \vee x \in B,$

and the inference of the desired conclusion from (1) is obvious. Q.E.D.

THEOREM 60. $\cup (A \cup B) = (\cup A) \cup (\cup B).$

PROOF.

$x \in \cup (A \cup B) \leftrightarrow (\exists C)(x \in C \,\&\, C \in A \cup B)$ by Theorem 55

$\qquad \leftrightarrow (\exists C)((x \in C \,\&\, C \in A) \vee (x \in C \,\&\, C \in B))$

                          by Theorem 20 and sentential logic

$\qquad \leftrightarrow (\exists C)(x \in C \,\&\, C \in A) \vee (\exists C)(x \in C \,\&\, C \in B)$

                          by quantifier logic*

$\qquad \leftrightarrow x \in \cup A \vee x \in \cup B$ by Theorem 55

$\qquad \leftrightarrow x \in \cup A \cup \cup B$ by Theorem 20. Q.E.D.

*Clearly, from $(\exists v) (P \vee Q)$ we may infer $(\exists v)P \vee (\exists v)Q$, and conversely.

THEOREM 61. $A \subseteq B \to \bigcup A \subseteq \bigcup B.$

PROOF.

$$x \in \bigcup A \leftrightarrow (\exists C)(x \in C \,\&\, C \in A) \qquad \text{by Theorem 55}$$
$$\to (\exists C)(x \in C \,\&\, C \in B) \qquad \begin{array}{l}\text{by hypothesis}\\ \text{of the theorem}\end{array}$$
$$\to x \in \bigcup B \qquad \text{by Theorem 55.}$$

Q.E.D.

THEOREM 62. $A \in B \to A \subseteq \bigcup B.$

THEOREM 63. $(\forall A)(A \in B \to A \subseteq C) \to \bigcup B \subseteq C.$

THEOREM 64. $(\forall A)(A \in B \to A \cap C = 0) \to (\bigcup B) \cap C = 0.$

THEOREM 65. $\bigcup \langle x,y \rangle = \{x,y\}.$

PROOF.

$$\bigcup \langle x,y \rangle = \bigcup \{\{x\},\{x,y\}\} \qquad \text{by definition of ordered pairs}$$
$$= \{x\} \cup \{x,y\} \qquad \text{by Theorem 59}$$
$$= \{x,y\} \qquad \text{Q.E.D.}$$

THEOREM 66. $\bigcup \bigcup \langle A,B \rangle = A \cup B.$

We now turn to the definition and properties of the *intersection* of a family of sets. The intuitive content of this notion should be clear from the discussions of union. If, as before,

$$A = \{\{1,2\}, \{2,3\}, \{4\}, \text{Jane Austen}\}$$

then

$$\cap A = 0,$$

since there is no number common to all the sets which are members of A. As a second example, if

$$B = \{\{1,2\}, \{2,3\}\}$$

then

$$\cap B = \{1,2\} \cap \{2,3\} = \{2\}.$$

The formal developments which follow require little comment.

DEFINITION 14. $\cap A = \{x : (\forall B)(B \in A \to x \in B)\}.$

There is no theorem related to Definition 14 in the way that Theorem 55 is related to Definition 13, that is, we cannot prove:

(1) $\qquad x \in \cap A \leftrightarrow (\forall B)(B \in A \to x \in B),$

and the reason is obvious. If A has no sets as members, then the right member of (1) is always true and every x must be a member of $\cap A$. But there is no set having every entity x as a member, a fact which was established by Theorem 41. What we are able to prove is the more restricted result:

Theorem 67. $x \in \cap A \leftrightarrow (\forall B)(B \in A \rightarrow x \in B)$ & $(\exists B)(B \in A)$.

proof [Necessity].   By hypothesis $x \in \cap A$.   Hence $\cap A \neq 0$.

Thus by virtue of Definition 14 and the general properties of definition by abstraction, we infer that

(1) $x \in \cap A \leftrightarrow (\forall B)(B \in A \rightarrow x \in B)$.

Now let us make the supposition that

(2) $-(\exists B)(B \in A)$.

Then vacuously it is true that

$$(\forall B)(B \in A \rightarrow x \in B),$$

from which we may infer:

(3) $(\forall B)(B \in A \rightarrow x \in B) \leftrightarrow x = x$.

Equivalences (1) and (3) yield that for every x

$$x \in \cap A \leftrightarrow x = x,$$

whence by Theorem 54

$$\{x : x \in \cap A\} = \{x : x = x\},$$

but by virtue of Theorem 48 the left-hand side is $\cap A$ and by Theorem 50 the right-hand side is the empty set, and so we infer

$$\cap A = 0$$

which contradicts the hypothesis that $x \in \cap A$ and proves our supposition (2) false.

[Sufficiency].   By hypothesis there is a set, say B^*, which is a member of A.   Hence we may apply the axiom schema of separation to obtain:

(4) $(\exists C)(\forall x)(x \in C \leftrightarrow x \in B^*$ & $(\forall B)(B \in A \rightarrow x \in B))$.

Since the fact that $x \in B^*$ follows from $B^* \in A$, and the other part of our hypothesis, namely that

$$(\forall B)(B \in A \rightarrow x \in B),$$

we infer from (4) that

(5) $(\exists C)(\forall x)(x \in C \leftrightarrow (\forall B)(B \in A \rightarrow x \in B))$.

That $x \in \cap A$ follows from (5), the definition of $\cap A$ and the defining conditions for definitions by abstraction. Q.E.D.

In this proof a square-bracket notation, which is often convenient in proving an equivalence, has been used. We consider the formula which is the *right* member of the equivalence as asserting a necessary and sufficient condition for the formula which is the left member to hold. Thus if we want to prove a theorem of the form $P \leftrightarrow Q$, we establish that Q is a *necessary* condition for P by assuming P and deriving Q. We establish that Q is a *sufficient* condition for P by deriving P from Q.

The order of development of this section is somewhat deceptive. The proof of Theorem 67 does not depend on the sum axiom, and thus the elementary theory of the $\cap$ operation could have preceded consideration of this axiom.

THEOREM 68. $\cap 0 = 0$.

PROOF. Suppose that $\cap 0 \neq 0$. Then there is an x in $\cap 0$, and by Theorem 67 there is a set $B \in 0$, which is absurd. Q.E.D.

It is worth remarking that in von Neumann set theory, which admits sets which are not members of any other set, that is, proper classes, the operation symbol '$\cap$' is defined in such a way that Theorem 68 is false. In fact, the theorem is then that

(1) $$\cap 0 = V,$$

where V is the universe, that is, the proper class which has as members everything which is a member of something. The radical difference between (1) and Theorem 68 emphasizes the slightly artificial character of any form of axiomatic set theory. Intuitively (1) may seem preferable to Theorem 68, but (1) entails the admission of proper classes, which appear rather bizarre from the standpoint of naive, intuitive set theory.

THEOREM 69. $\cap \{0\} = 0$.

PROOF. Suppose there is an element x in $\cap \{0\}$. Then by virtue of Definition 14 $x \in 0$, which is absurd. Q.E.D.

Like the previous two theorems the next four theorems are concerned with the intersection of extremely simple families of sets. Two of the proofs are omitted.

THEOREM 70. $\cap \{A\} = A$.

PROOF. If

$$x \in \cap \{A\},$$

then since $A \in \{A\}$, by Definition 14

$$x \in A.$$

On the other hand, if $x \in A$, then since for every B in $\{A\}$, $B = A$, by Theorem 67,

$$x \in \cap \{A\}. \qquad \text{Q.E.D.}$$

THEOREM 71. $\cap \{A, B\} = A \cap B$.

THEOREM 72. $\cap \langle x, y \rangle = \{x\}$.

THEOREM 73. $\cap \cap \langle A, B \rangle = A$.

PROOF.   By virtue of Theorem 72

$$\cap \langle A, B \rangle = \{A\},$$

and by virtue of Theorem 70

$$\cap \{A\} = A. \qquad \text{Q.E.D.}$$

Five general implications concerning intersections of families of sets are next.

THEOREM 74. $A \subseteq B \,\&\, (\exists C)(C \in A) \to \cap B \subseteq \cap A$.

PROOF.   Let x be an arbitrary element of $\cap B$. Then for every $C \in B$, we must have:

$$x \in C,$$

but the hypothesis of the theorem assures us that if $C \in A$ then $C \in B$. Whence for every $C \in A$, we must have: $x \in C$, and thus $x \in \cap A$, the desired result.   Q.E.D.

Exact explanation of why the condition that $(\exists C)(C \in A)$ is needed in the hypothesis of Theorem 74 is left as an exercise.

THEOREM 75. $A \in B \to \cap B \subseteq A$.

THEOREM 76. $A \in B \,\&\, A \subseteq C \to \cap B \subseteq C$.

THEOREM 77. $A \in B \,\&\, A \cap C = 0 \to (\cap B) \cap C = 0$.

THEOREM 78. $(\exists C)(C \in A) \,\&\, (\exists D)(D \in B) \to \cap (A \cup B) = (\cap A) \cap (\cap B)$.

PROOF.

$x \in \cap(A \cup B) \leftrightarrow (\forall C)(C \in A \cup B \to x \in C)$ by Theorem 67 and Hypothesis of Theorem

$\leftrightarrow (\forall C)(C \in A \lor C \in B \to x \in C)$ by Theorem 20

$$\leftrightarrow (\forall C)[(C \in A \rightarrow x \in C) \; \& \; (C \in B \rightarrow x \in C)]$$

by sentential logic

$$\leftrightarrow (\forall C)(C \in A \rightarrow x \in C) \; \& \; (\forall C)(C \in B \rightarrow x \in C)$$

by quantifier logic*

$$\leftrightarrow x \in \cap A \; \& \; x \in \cap B$$

by Theorem 67 and Hypothesis of Theorem

$$\leftrightarrow x \in (\cap A) \cap (\cap B)$$

by Theorem 12.

The exact formulation of Theorem 78 is sensitive to the form of axiomatic set theory being used. If proper classes are admitted and $\cap 0 = V$, then the theorem is stated unconditionally:

$$\cap (A \cup B) = (\cap A) \cap (\cap B).$$

If Zermelo set theory *without* individuals is the framework, then the formulation is simply:

(1) $\qquad A \neq 0 \; \& \; B \neq 0 \rightarrow \cap (A \cup B) = (\cap A) \cap (\cap B).$

Zermelo set theory *with* individuals requires the formulation given in the theorem. The incorrectness of (1) for our framework of development is seen by taking:

$$A = \{\text{Euler}\} \neq 0$$
$$B = \{\{\text{Euler}\}\} \neq 0.$$

Then

$$\cap A = 0$$

and hence

$$(\cap A) \cap (\cap B) = 0,$$

but

$$\cap (A \cup B) = \{\text{Euler}\} \neq 0.$$

The next group of theorems involves both intersection and union of a family of sets.

---

*Clearly from $(\forall v)(P \; \& \; Q)$ we may infer $(\forall v)P \; \& \; (\forall v)Q$, and conversely.

THEOREM 79. $\cap A \subseteq \cup A$.

PROOF. If $x \in \cap A$, then by Theorem 67

(1) $(\forall B)(B \in A \rightarrow x \in B) \; \& \; (\exists B)(B \in A)$.

It follows from (1) that

$$(\exists B)(B \in A \; \& \; x \in B),$$

whence by Theorem 55, $x \in \cup A$. Q.E.D.

THEOREM 80. $\cup \cap \langle A,B \rangle = A$.

THEOREM 81. $\cap \cup \langle A,B \rangle = A \cap B$.

In many mathematical contexts in place of $\cup A$ and $\cap A$ the notation

(1) $\underset{B \in A}{\cup} B$

and

(2) $\underset{B \in A}{\cap} B$

is often seen. In fact, notation more flexible than (1) and (2) is convenient in later chapters for the development of ordinal number theory. In introducing at this point the appropriate definitional schema, we use the expression '$\tau(x)$' for a term schema just as we have used '$\varphi(x)$' for a formula schema.

DEFINITION SCHEMA 15.

(a) $\underset{x \in A}{\cup} \tau(x) = \cup \{y \colon (\exists x)(y = \tau(x) \; \& \; x \in A)\}$

(b) $\underset{x \in A}{\cap} \tau(x) = \cap \{y \colon (\exists x)(y = \tau(x) \; \& \; x \in A)\}$.

Thus, if $A = \{1,2,3\}$ and $\tau(x) = \{x\} \cup \{4\}$, then

$$\underset{x \in A}{\cup} \tau(x) = \cup \{\{1,4\}, \{2,4\}, \{3,4\}\}$$

$$= \{1,2,3,4\}$$

and

$$\underset{x \in A}{\cap} \tau(x) = \{4\}.$$

Still other notational devices for union and intersection of families of sets are often useful, but will not be formally introduced. For instance,

$$\underset{\varphi(x)}{\cup} x = \cup \{x \colon \varphi(x)\}.$$

From a logical standpoint there is a sharp difference between Definitions 13 and 14 on the one hand and Definition 15 on the other. The first two introduce operation symbols, whereas Definition 15 introduces an operator which provides a new way of binding variables, and in this respect stands together with Definitions 11 and 12.

We state without proof some theorems concerning the notions introduced by Definition 15. Note that these theorems, like Theorem 78, assert important general distributive laws. Some additional results are given in the exercises.

THEOREM 82. $\bigcup\limits_{x \in A} x = \bigcup A.$

THEOREM 83. $\bigcap\limits_{x \in A} x = \bigcap A.$

THEOREM 84. $A \cap \bigcup B = \bigcup\limits_{C \in B} (A \cap C).$

THEOREM 85. $(\exists D)(D \in B) \rightarrow A \cup \bigcap B = \bigcap\limits_{C \in B} (A \cup C).$

Finally, we conclude this section by showing that the union axiom is redundant. Since this theorem is not about sets but about our particular axioms for set theory we list it as a metatheorem, that is, as a metamathematical theorem.

METATHEOREM 1. *The union axiom is derivable from the axiom of extensionality, the pairing axiom, and the sum axiom.*

PROOF. Given any two sets A and B, by the pairing axiom, we have the set

$$\{A, B\}.$$

Now

$$x \in \bigcup\{A, B\} \leftrightarrow (\exists D)(D \in \{A, B\} \,\&\, x \in D)$$

by Theorem 55

$$\leftrightarrow (\exists D)((D = A \lor D = B) \,\&\, x \in D)$$

by Theorem 43

$$\leftrightarrow x \in A \lor x \in B \qquad \text{by quantifier logic.}$$

From the above equivalences it is a simple matter of quantifier logic to derive that

$$(\exists C)(\forall x)(x \in C \leftrightarrow x \in A \lor x \in B),$$

which is precisely the union axiom. Q.E.D.

In connection with the above proof, it is easy to check that Theorems 43 and 55 depend on no more than the three axioms mentioned. Precise identification of the points where the axiom of extensionality is needed is left as an exercise.

EXERCISES

1. Given that

$$A = \{\{1,2\}, \{2,0\}, \{1,3\}\}$$

find $\cup A$, $\cap A$, $\cap \cup A$.

2. Given that

$$A = \{\{\{1,2\}, \{1\}\}, \{\{1,0\}\}\}$$

find $\cup A$, $\cap A$, $\cup \cup A$, $\cap \cap A$, $\cup \cap A$, $\cap \cup A$.

3. Give a specific set A which will serve as a counterexample to the general assertion that

$$\cap A = 0 \rightarrow A = 0 \lor A = \{0\}.$$

4. Give specific sets A and B which will yield a counterexample to the general assertion that

$$\cap A \cap \cap B = \cap (A \cap B).$$

5. Give a definition of $\cup A$ by means of an equivalence and without use of the abstraction notation.

6. In Zermelo's original paper [1908], he defined '$\cap$' in such a way that if A has an individual as an element, $\cap A = 0$. Reformulate Definition 14 to conform to Zermelo's, and show by means of counterexamples which of the theorems in this section do not hold when this revised definition is used.

7. Prove Theorem 58.

8. Prove Theorems 62, 63, and 64.

9. Prove Theorem 66.

10. Prove Theorems 71 and 72.

11. With respect to the existential part of the hypothesis of Theorem 74, show by an example that if it is omitted the resulting statement is not a theorem.

12. Prove Theorems 75, 76, and 77.

13. Prove Theorems 80 and 81.

14. Prove that

$$0 \in A \rightarrow \cap A = 0.$$

15. Prove that

$$(\forall C)(C \in A \rightarrow (\exists D)(D \in B \,\&\, C \subseteq D)) \rightarrow \cup A \subseteq \cup B.$$

16. Prove Theorems 82 and 83.

17. Prove Theorems 84 and 85 (use power set axiom, p. 47).

18. Prove that $\displaystyle\bigcap_{x \in A} (\{x\} \sim B) = \bigcap_{x \in A} \{x\} \sim B.$

19. Explain at what points the axiom of extensionality is needed in the proof of Metatheorem 1.

§ 2.7 **Power Set Axiom.** In this section we are concerned with the notion of the set of all subsets of a given set. This set is called the *power set* of a given set. The name 'power set' has its origin in the fact that if a set A has n elements, then its power set (in symbols: $\mathcal{P}A$) has 2^n elements. As an illustration of the notion, if

$$A = \{1,2\}$$

then

$$\mathcal{P}A = \{0, \{1\}, \{2\}, A\}.$$

It should be intuitively clear that, as in this example, the empty set is a member of the power set of any set; moreover, any set is a member of its own power set.

The appropriate formal definition should be obvious.

DEFINITION 16. $\mathcal{P}A = \{B : B \subseteq A\}$.

This definition is axiom-free in the same sense that Definitions 13 and 14 are, but in order to prove the desired theorem concerning $\mathcal{P}A$, the *power set axiom* guaranteeing the existence of the intuitively appropriate set is needed:

$$(\exists B)(\forall C)(C \in B \leftrightarrow C \subseteq A).$$

It is worth noting that we could have taken the weaker formulation '$(\exists B)(\forall C)(C \subseteq A \rightarrow C \in B)$' and then used the axiom schema of separation to get the present axiom. We may immediately prove:

THEOREM 86. $B \in \mathcal{P}A \leftrightarrow B \subseteq A$.

PROOF.   Use Definition 16, power set axiom and properties of definition by abstraction.

THEOREM 87. $A \in \mathcal{P}A$.

PROOF.   By Theorem 5

$$A \subseteq A,$$

whence by Theorem 86 we obtain the desired result.   Q.E.D.

THEOREM 88. $0 \in \mathcal{P}A$.

THEOREM 89. $\mathcal{P}0 = \{0\}$.

PROOF.   Since $0 \subseteq 0$,

$$0 \in \mathcal{P}0.$$

Moreover, if $A \in \mathcal{P}0$, then by Theorem 86

$$A \subseteq 0,$$

but then by Theorem 4

$$A = 0. \qquad\qquad\qquad\qquad \text{Q.E.D.}$$

THEOREM 90.　$\mathcal{P}\mathcal{P}0 = \{0, \{0\}\}$.

There are only four further theorems concerning power sets which we wish to state in this section.

THEOREM 91.　$A \subseteq B \leftrightarrow \mathcal{P}A \subseteq \mathcal{P}B$.

PROOF.　[Necessity].　If $C \in \mathcal{P}A$ then by Theorem 86

$$C \subseteq A,$$

whence by our hypothesis

$$C \subseteq B,$$

and thus by virtue of Theorem 86 again

$$C \in \mathcal{P}B.$$

[Sufficiency].　By virtue of Theorem 87, $A \in \mathcal{P}A$, and thus by our hypothesis that $\mathcal{P}A \subseteq \mathcal{P}B$,

$$A \in \mathcal{P}B,$$

but then by Theorem 86 $A \subseteq B$.　Q.E.D.

THEOREM 92.　$(\mathcal{P}A) \cup (\mathcal{P}B) \subseteq \mathcal{P}(A \cup B)$.

PROOF.

$$C \in (\mathcal{P}A) \cup (\mathcal{P}B) \leftrightarrow C \in \mathcal{P}A \vee C \in \mathcal{P}B$$
$$\leftrightarrow C \subseteq A \vee C \subseteq B$$
$$\rightarrow C \subseteq A \cup B$$
$$\rightarrow C \in \mathcal{P}(A \cup B).\qquad\qquad \text{Q.E.D.}$$

The justification of the steps in this proof should be obvious.

THEOREM 93.　$\mathcal{P}(A \cap B) = (\mathcal{P}A) \cap (\mathcal{P}B)$.

THEOREM 94.　$\mathcal{P}(A \sim B) \subseteq ((\mathcal{P}A) \sim (\mathcal{P}B)) \cup \{0\}$.

EXERCISES

1.　Find

$$\mathcal{P}\{\text{Archimedes}\},$$
$$\mathcal{P}\mathcal{P}\{\text{Archimedes}\},$$
$$\mathcal{P}\{\{\text{Archimedes, Newton}\}, 0\}.$$

2.　Find

$$\mathcal{P}\mathcal{P}\mathcal{P}0.$$

3.　Prove Theorems 88 and 90.

4. Prove Theorem 93.

5. Prove Theorem 94.

6. Give counterexamples to show that it is not always the case that

   (a) $(\mathcal{P}A) \cup (\mathcal{P}B) = \mathcal{P}(A \cup B)$

   (b) $\mathcal{P}(A \sim B) = (\mathcal{P}A) \sim (\mathcal{P}B)$.

§ 2.8 Cartesian Product of Sets.   The Cartesian product of two sets A and B (in symbols: $A \times B$) is the set of all ordered pairs $\langle x, y \rangle$ such that $x \in A$ and $y \in B$. For example, if

$$A = \{1,2\}$$

$$B = \{\text{Archimedes, Eudoxus}\}$$

then

$$A \times B = \{\langle 1, \text{Archimedes}\rangle, \langle 1, \text{Eudoxus}\rangle, \langle 2, \text{Archimedes}\rangle \\ \langle 2, \text{Eudoxus}\rangle\}.$$

Formally, we have:

DEFINITION 17. $A \times B = \{\langle x,y \rangle : x \in A \ \& \ y \in \mathrm{B}\}$.

In order to prove the standard theorems about Cartesian products, we must show that the intuitively appropriate set exists. The proof of this fact depends in an essential way on the power set axiom. The crucial idea of the proof is that if

$$x = \langle y,z \rangle,$$

$$y \in A \text{ and } z \in B$$

then

$$x \in \mathcal{P}\mathcal{P}(A \cup B).$$

THEOREM 95. $(\exists C)(\forall x)(x \in C \leftrightarrow (\exists y)(\exists z)(y \in A \ \& \ z \in B \ \& \\ x = \langle y,z \rangle))$.

PROOF.   By virtue of the axiom schema of separation

(1) $(\exists C)(\forall x)(x \in C \leftrightarrow x \in \mathcal{P}\mathcal{P}(A \cup B) \ \& \\ (\exists y)(\exists z)(y \in A \ \& \ z \in B \ \& \ x = \langle y,z \rangle))$.

Since the theorem is just (1) without '$x \in \mathcal{P}\mathcal{P}(A \cup B)$', our task is to show that the equivalence given in (1) still holds when this clause is eliminated. Given (1) it follows at once that

(2) $x \in C$

implies

(3) $(\exists y)(\exists z)(y \in A \ \& \ z \in B \ \& \ x = \langle y,z \rangle)$.

To establish the converse implication it will suffice to show that (3) implies

(4) $$x \in \mathscr{P}\mathscr{P}(A \cup B),$$

since by virtue of (1) it will then be obvious that (3) implies (2).

Thus we need show only that (3) implies (4). Now by (3) and the definition of ordered pairs

$$x = \{\{y\}, \{y,z\}\},$$

and since by hypothesis $y \in A$ and $z \in B$, we have:

$$\{y\} \subseteq A \cup B,$$

and

$$\{y,z\} \subseteq A \cup B,$$

whence by virtue of Theorem 86

$$\{y\} \in \mathscr{P}(A \cup B)$$

and

$$\{y, z\} \in \mathscr{P}(A \cup B).$$

Thus

$$\{\{y\}, \{y, z\}\} \subseteq \mathscr{P}(A \cup B),$$

that is,

$$x \subseteq \mathscr{P}(A \cup B),$$

but by virtue of Theorem 86 again, we then have:

$$x \in \mathscr{P}\mathscr{P}(A \cup B),$$

which is what we desired to prove.   Q.E.D.

We have then almost immediately the following two useful theorems.

THEOREM 96. $x \in A \times B \leftrightarrow (\exists y)(\exists z)(y \in A \ \& \ z \in B \ \& \ x = \langle y,z \rangle)$.

THEOREM 97. $\langle x,y \rangle \in A \times B \leftrightarrow x \in A \ \& \ y \in B$.

We next turn to a number of theorems whose intuitive content is obvious. Several of the proofs are omitted and left as exercises.

THEOREM 98. $A \times B = 0 \leftrightarrow A = 0 \lor B = 0$.

PROOF [Necessity].   We use an indirect argument. By hypothesis $A \times B = 0$. Suppose now that

$$A \neq 0 \ \& \ B \neq 0.$$

Then by Theorem 2

$$(\exists y)(y \in A) \ \& \ (\exists z)(z \in B),$$

and thus by Theorem 97

$$\langle y,z \rangle \in A \times B,$$

which contradicts the hypothesis and proves our supposition false.

[Sufficiency]. From the condition that $A = 0$ or $B = 0$ and Theorem 2, we infer

(1) $$-(\exists y)(y \in A) \lor -(\exists z)(z \in B),$$

and it follows from (1) that

$$-(\exists y)(\exists z)(y \in A \ \& \ z \in B \ \& \ x = \langle y,z \rangle),$$

and thus by Theorem 96, for every x

$$x \notin A \times B.$$

Hence by Theorem 2

$$A \times B = 0. \hspace{3cm} \text{Q.E.D.}$$

THEOREM 99. $A \times B = B \times A \leftrightarrow (A = 0 \lor B = 0 \lor A = B)$.

PROOF [Necessity]. Suppose that $A \neq 0$, $B \neq 0$ and $A \neq B$, i.e., suppose the condition does not hold. Since $A \neq B$, there is an x such that either $x \in A \ \& \ x \notin B$ or $x \notin A \ \& \ x \in B$. For definiteness, let us assume the first alternative holds, and let y be an element of B (there are such elements, since $B \neq 0$). Then by Theorem 97

$$\langle x,y \rangle \in A \times B,$$

and thus from the hypothesis that $A \times B = B \times A$, we have:

$$\langle x,y \rangle \in B \times A$$

but by virtue of Theorem 97 again

$$x \in B,$$

which contradicts our assumption that $x \notin B$.

[Sufficiency]. Of the three possibilities we may use Theorem 98 to combine two; namely, from $A = 0 \lor B = 0$ we infer that

$$A \times B = 0 = B \times A.$$

Assume now the third possibility: $A = B$. Then since it is a truth of logic that

$$A \times A = A \times A$$

we have at once that

$$A \times B = B \times A. \hspace{3cm} \text{Q.E.D.}$$

THEOREM 100. $A \neq 0 \ \& \ A \times B \subseteq A \times C \to B \subseteq C.$

PROOF. Since by hypothesis $A \neq 0$, let

$$x \in A \ \& \ y \in B.$$

Then by Theorem 97

$$\langle x,y \rangle \in A \times B,$$

and thus by hypothesis

$$\langle x,y \rangle \in A \times C.$$

Hence by use of Theorem 97 again

(1) $$\hspace{5cm} y \in C.$$

Since y is an arbitrary element of B, (1) establishes that $B \subseteq C$.   Q.E.D.
The proof of the next theorem is left as an exercise.

THEOREM 101. $B \subseteq C \to A \times B \subseteq A \times C.$

The next three theorems state three distributive laws for the operation of forming the Cartesian product of two sets. Only the first one is proved here.

THEOREM. 102. $A \times (B \cap C) = (A \times B) \cap (A \times C).$

PROOF.

$$\langle x,y \rangle \in A \times (B \cap C) \leftrightarrow x \in A \ \& \ y \in B \cap C \hspace{1.5cm} \text{by Theorem 97}$$

$$\leftrightarrow x \in A \ \& \ y \in B \ \& \ y \in C \hspace{1.5cm} \text{by Theorem 12}$$

$$\leftrightarrow x \in A \ \& \ y \in B \ \& \ x \in A \ \& \ y \in C$$

$$\text{by sentential logic}$$

$$\leftrightarrow \langle x,y \rangle \in A \times B \ \& \ \langle x,y \rangle \in A \times C$$

$$\text{by Theorem 97}$$

$$\leftrightarrow \langle x,y \rangle \in (A \times B) \cap (A \times C)$$

$$\text{by Theorem 12.}$$

$$\text{Q.E.D.}$$

THEOREM 103. $A \times (B \cup C) = (A \times B) \cup (A \times C).$

THEOREM 104. $A \times (B \sim C) = (A \times B) \sim (A \times C).$

EXERCISES

1. Prove Theorems 96 and 97.
2. Prove Theorem 101.
3. Prove Theorems 103 and 104.
4. Give a simple counterexample to show that in general it is not the case that

$$A \cup (B \times C) = (A \times B) \cup (A \times C).$$

5. Is the Cartesian product operation associative? If so, prove it. If not, give a counterexample.
6. Prove that

$$A \times \cap B = \bigcap_{C \in B} (A \times C).$$

§ 2.9 Axiom of Regularity. It is difficult to think of a set which might reasonably be regarded as a member of itself. Certainly the set of all men, for example, is not a man and is therefore not a member of itself. Perhaps it might be argued that in intuitive set theory the set of all abstract objects or the set of all sets should provide an example of a set which is a member of itself, but as we saw in the first chapter, the set of all sets is itself a paradoxical concept.

These remarks suggest we take as an axiom:

$$(1) \qquad\qquad A \notin A.$$

However, the assumption of (1) would not prohibit the counterintuitive situation of there being distinct sets A and B such that

$$(2) \qquad\qquad A \in B \,\&\, B \in A.$$

(If you do not believe (2) is counterintuitive, try to give a simple example of sets A and B satisfying (2).) Furthermore, if we took (2) as an axiom, longer counterintuitive cycles of membership would not be ruled out — like the existence of distinct sets A, B, and C such that

$$(3) \qquad\qquad A \in B \,\&\, B \in C \,\&\, C \in A.$$

We prevent such cycles of any length n by adopting an axiom which is, on the assumption of our other axioms, including the axiom of choice, equivalent to the non-existence of infinite descending sequences of sets (i.e., $A_{i+1} \in A_i$). The form of the axiom which we adopt, the *axiom of regularity*, is due to Zermelo [1930], although an essentially equivalent but more complicated axiom was given earlier in von Neumann [1929, p. 231];*

$$A \neq 0 \to (\exists x)[x \in A \,\&\, (\forall y)(y \in x \to y \notin A)].$$

*The essential idea was formulated even earlier in von Neumann [1925, p. 239] and prior to that in Mirimanoff [1917].

This axiom was called by Zermelo the *Axiom der Fundierung*. Intuitively it says that given any non-empty set A there is a member x of A such that the intersection of A and x is empty. The part '$(\forall y)(y \in x \to y \notin A)$' which expresses that the intersection of A and x is empty has not been replaced by the simpler appearing formula '$A \cap x = 0$' because of the conditional definition of intersection; for if x is an individual, the definition assigns no intuitive meaning to the intersection of x and any other object. When it is clear that x must be a set, we use the simpler formula in proofs.

We now use the axiom of regularity to prove (1) and the negation of (2) as theorems.

THEOREM 105. $A \notin A$.

PROOF. Suppose that A is a set such that $A \in A$. Since $A \in \{A\}$, we then have

(1) $$A \in \{A\} \cap A.$$

By virtue of the axiom of regularity there is an x in $\{A\}$ such that

$$\{A\} \cap x = 0,$$

but since $\{A\}$ is a unit set,

$$x = A,$$

and thus

$$\{A\} \cap A = 0,$$

which contradicts (1). Q.E.D.

THEOREM 106. $-(A \in B \ \& \ B \in A)$.

PROOF. Suppose that $A \in B \ \& \ B \in A$. Then

(1) $$A \in \{A,B\} \cap B \text{ and } B \in \{A,B\} \cap A.$$

By the axiom of regularity there is an x in $\{A,B\}$ such that

$$\{A,B\} \cap x = 0$$

and by Theorem 43

$$x = A \quad \text{or} \quad x = B.$$

Hence

$$\{A,B\} \cap A = 0 \quad \text{or} \quad \{A,B\} \cap B = 0,$$

which contradicts (1). Q.E.D.

The proof of Theorem 106 proceeded exactly as did that of the previous theorem. Proof of the impossibility of a cycle of three or more sets proceeds similarly.

As an example of the kind of theorem for which the axiom of regularity is essential, we may prove a theorem about Cartesian products which may seem intuitively obvious, but which cannot be proved on the basis of only the axioms introduced earlier.

THEOREM 107. $A \subseteq A \times A \rightarrow A = 0$.

PROOF. Since by hypothesis A is a subset of $A \times A$, from the definition of Cartesian product we know that if $z \in A$ then there are elements x and y such that

$$(1) \qquad z = \langle x,y \rangle = \{\{x\}, \{x,y\}\}$$

and

$$(2) \qquad x \in A \ \& \ y \in A.$$

Suppose now contrary to the theorem that $A \neq 0$. Let us apply the axiom of regularity to $A \cup \cup A$; whence there is a non-empty set C such that

$$C \in A \cup \cup A$$

and

$$(3) \qquad C \cap (A \cup \cup A) = 0.$$

That C must be a non-empty set, and not the empty set or an individual, follows from (1) — the elements of both A and $\cup A$ must be non-empty sets. Suppose now that $C \in A$. Then by Theorem 62, $C \subseteq \cup A$ and since C is non-empty, we must have

$$C \cap \cup A \neq 0,$$

which contradicts (3). Thus C must be in $\cup A$, but on the basis of (1), there are elements x and y such that

$$C = \{x\} \vee C = \{x,y\}$$

and on the basis of (2), $x,y \in A$, whence in either case

$$C \cap A \neq 0,$$

which also contradicts (3), and proves that our supposition that $A \neq 0$ is false.   Q.E.D.

Even though the axiom of regularity has very natural consequences and imposes, as Zermelo remarked in his 1930 paper, a condition which will be satisfied in all practical applications, it is possible to construct systems of set theory which contradict this axiom. Two examples are Lesniewski's system of ontology (for a good account see Slupecki [1955]) and the system of Quine [1940].

EXERCISES

1. Prove that for all sets A, B, and C it is not the case that

$$A \in B \,\&\, B \in C \,\&\, C \in A.$$

2. Prove that if $A = A \times B$ then $A = 0$.

3. Prove that if $A \times B \neq 0$ then there is a C in $A \times B$ such that $(\bigcup C) \cap (A \times B) = 0$.

4. Prove an analogue of Theorem 46 for the following definition of ordered pairs:

$$\langle x,y \rangle = \{x, \{x,y\}\}.$$

§ **2.10 Summary of Axioms.** For convenient subsequent reference we summarize here the six non-redundant axioms introduced in this chapter. The union axiom is omitted because it was shown in §2.6 that it follows from the axiom of extensionality, the pairing axiom, and the sum axiom. These six axioms suffice for all developments in Chapter 3, which is concerned with relations and functions.

Axiom of Extensionality:

$$(\forall x)(x \in A \leftrightarrow x \in B) \to A = B.$$

Axiom Schema of Separation:

$$(\exists B)(\forall x)(x \in B \leftrightarrow x \in A \,\&\, \varphi(x)).$$

Pairing Axiom:

$$(\exists A)(\forall z)(z \in A \leftrightarrow z = x \vee z = y).$$

Sum Axiom:

$$(\exists C)(\forall x)(x \in C \leftrightarrow (\exists B)(x \in B \,\&\, B \in A)).$$

Power Set Axiom:

$$(\exists B)(\forall C)(C \in B \leftrightarrow C \subseteq A).$$

Axiom of Regularity:

$$A \neq 0 \to (\exists x)[x \in A \,\&\, (\forall y)(y \in x \to y \notin A)].$$

CHAPTER 3

RELATIONS AND FUNCTIONS

---

§ 3.1 Operations on Binary Relations. In everyday contexts we frequently speak of *relations* which hold between two, or among several, things. Thus we may say that Augustus stood in the relation of stepfather to Tiberius, or that the relation of betweenness holds among three points. When we refer to relations in ordinary contexts we insist that there be some intuitive description of the sort of connection existing between items which stand in a given relation. Fortunately this vague idea of intuitive connectedness may be dispensed with in formal contexts and a relation may be defined simply as a set of ordered pairs. We shall in this chapter be concerned almost entirely with the theory of binary relations, that is, relations which hold between two things. Moreover, as we shall see, the theory of *n*-place relations may be constructed within the theory of binary relations. Consequently we omit the modifying adjective 'binary' in the formal definition.*

DEFINITION 1. *A is a relation* $\leftrightarrow (\forall x)(x \in A \rightarrow (\exists y)(\exists z)(x = \langle y, z \rangle))$.

It is interesting to note that this is our first definition of a one-place relation symbol since Definition 1 of Chapter 2, which characterized the property of being a set. A natural idea would be that subsequent definitions concerning relations must, like the definitions following the definition of sets, be largely conditional in form. However, this is not the case, and nearly all of what follows applies to arbitrary sets, not just those special sets which happen to be relations.

The subsumption of *n*-place (or *n*-ary) relations under this definition may be exemplified by considering ternary, that is, three-place relations.

---

*Definitions and theorems are numbered anew in each chapter. A reference to a definition or theorem without explicit mention of a chapter is to a definition or theorem in the same chapter as the reference.

A set A is a ternary relation if and only if A is a relation and

$$(\forall x)(x \in A \rightarrow (\exists y)(\exists z)(\exists w)(x = \langle\langle y, z\rangle, w\rangle)).$$

On the other hand, notice that not for every intuitive relation which occurs in set theory is there a corresponding set of ordered pairs. For instance, there is no set corresponding to the inclusion relation between sets. In von Neumann set theory there is a proper class which is the inclusion relation between sets, but there is no proper class corresponding to the inclusion relation between proper classes.

We begin systematic developments with three simple theorems, after first introducing the useful notation: $x \, A \, y$.

DEFINITION 2. $x \, A \, y \leftrightarrow \langle x, y\rangle \in A$.

THEOREM 1. 0 *is a relation.*

PROOF.   Immediate from the definition of relations, since the empty set has no members.

THEOREM 2. R *is a relation* $\&$ $S \subseteq R \rightarrow S$ *is a relation.*

PROOF.   Let x be an arbitrary element of S. Then by hypothesis

$$x \in R,$$

whence, also by our hypothesis, there is a y and a z such that

$$x = \langle y, z\rangle.$$

Hence, according to Definition 1, S is a relation.   Q.E.D.

THEOREM 3. R *and* S *are relations* $\rightarrow R \cap S$, $R \cup S$ *and* $R \sim S$ *are relations.*

Use of the variables 'R' and 'S' in the last two theorems is no formal innovation, for all capital italic letters are set variables; it is meant to be merely suggestive of the fact that we are intuitively thinking about those sets which are relations, although the theorems hold for arbitrary sets.

If R is a relation then the *domain* of R (in symbols: $\mathfrak{D}R$) is the set of all things x such that, for some y, $\langle x, y\rangle \in R$. Thus if

$$R_1 = \{\langle 0, 1\rangle, \langle 2, 3\rangle\}$$

then

$$\mathfrak{D}R_1 = \{0, 2\}.$$

The *range* of R (in symbols: $\mathfrak{R}R$) is the set of all things y such that, for some x, $\langle x, y\rangle \in R$. Thus

$$\mathfrak{R}R_1 = \{1, 3\}.$$

The range of a relation is also called the *counterdomain* or *converse domain*. The *field* of a relation R (in symbols: $\mathfrak{F}R$) is the union of its domain and range. For example,

$$\mathfrak{F}R_1 = \{0, 1, 2, 3\}.$$

In the obvious formal developments connected with the three concepts of domain, range, and field the only difficult problem is to show that the intuitively appropriate set exists. As usual the definitions themselves are axiom-free.

DEFINITION 3. $\mathfrak{D}A = \{x : (\exists y)(x\,A\,y)\}$.

That $\mathfrak{D}A$ is the appropriate set is confirmed by the following theorem.

THEOREM 4. $x \in \mathfrak{D}A \leftrightarrow (\exists y)(x\,A\,y)$.

PROOF.   By virtue of the axiom schema of separation

(1) $(\exists B)(\forall x)(x \in B \leftrightarrow x \in \bigcup\bigcup A\ \&\ (\exists y)(x\,A\,y))$.

We want to establish the equivalence obtained from (1) by omitting:

(2) $x \in \bigcup\bigcup A$.

Consequently, we need to show that (2) is implied by the assertion that there is a y such that

(3) $x\,A\,y$.

The following chain of implications serves the purpose. By Definition 2 we have from (3):

$$\langle x, y \rangle \in A,$$

whence by virtue of the definition of ordered pairs

$$\{\{x\}, \{x, y\}\} \in A.$$

Thus by Theorem 55 of Chapter 2

$$\{x\} \in \bigcup A,$$

and by virtue of Theorem 55 again

$$x \in \bigcup\bigcup A.$$

It thus follows easily from (1) that

(4) $(\exists B)(\forall x)(x \in B \leftrightarrow (\exists y)(x\,A\,y))$.

The remainder of the proof simply requires routine manipulation of definition by abstraction. Perhaps it will be useful again to put in the details. By appropriate substitution in Definition Schema 11 of Chapter 2 we obtain:

(5) $\mathfrak{D}A = \{x : (\exists y)(x \, A \, y)\} \leftrightarrow [(\forall x)(x \in \mathfrak{D}A \leftrightarrow (\exists y)((x \, A \, y)) \lor$
$[-(\exists B)(\forall x)(x \in B \leftrightarrow (\exists y)(x \, A \, y)) \, \& \, \mathfrak{D}A = 0]].$

By virtue of Definition 3 above we obtain at once from (5)

(6) $(\forall x)(x \in \mathfrak{D}A \leftrightarrow (\exists y)(x \, A \, y)) \lor$
$[-(\exists B)(\forall x)(x \in B \leftrightarrow (\exists y)(x \, A \, y)) \, \& \, \mathfrak{D}A = 0],$

and from (4) and (6), we immediately conclude that our theorem holds.

Q.E.D.

Notice that in spite of the formidable appearance of (5) and (6), the inference of the theorem from (4), (5), (6) and Definition 3 involves only sentential logic.

THEOREM 5. $\mathfrak{D}(A \cup B) = \mathfrak{D}A \cup \mathfrak{D}B.$

PROOF.

$x \in \mathfrak{D}(A \cup B) \leftrightarrow (\exists y)(x \, A \cup B \, y)$	by Theorem 4
$\leftrightarrow (\exists y)(x \, A \, y \lor x \, B \, y)$	by Theorem 20 of Chapter 2
$\leftrightarrow (\exists y)(x \, A \, y) \lor (\exists y)(x \, B \, y)$	by quantifier logic
$\leftrightarrow x \in \mathfrak{D}A \lor x \in \mathfrak{D}B$	by Theorem 4
$\leftrightarrow x \in \mathfrak{D}A \cup \mathfrak{D}B$	by Theorem 20 of Chapter 2

Q.E.D.

Two similar theorems relating domains and intersections and differences are stated without proof.

THEOREM 6. $\mathfrak{D}(A \cap B) \subseteq \mathfrak{D}A \cap \mathfrak{D}B.$

THEOREM 7. $\mathfrak{D}A \sim \mathfrak{D}B \subseteq \mathfrak{D}(A \sim B).$

The notion of *range* may be defined symmetrically to that of domain.

DEFINITION 4. $\mathfrak{R}A = \{y : (\exists x)(x \, A \, y)\}.$

Since the theorems on the range operation parallel those on the domain operation, proofs are omitted. Moreover, the obvious analogue of Theorem 4 is not stated.

THEOREM 8. $\mathfrak{R}(A \cup B) = \mathfrak{R}A \cup \mathfrak{R}B.$

THEOREM 9. $\mathfrak{R}(A \cap B) \subseteq \mathfrak{R}A \cap \mathfrak{R}B.$

THEOREM 10. $\mathcal{R}A \sim \mathcal{R}B \subseteq \mathcal{R}(A \sim B)$.

The notion of the *field* of a set is defined as expected.

DEFINITION 5. $\mathcal{F}A = \mathcal{D}A \cup \mathcal{R}A$.

At the moment we shall prove no theorems about the field operation $\mathcal{F}$, but we shall subsequently make use of it.

We now turn to the important notion of *converse operation*. As in the case of the three operations just introduced, the definition applies to arbitrary sets, not merely to relations. The converse of a relation R (in symbols: $\breve{R}$) is the relation such that for all x and y, $x \breve{R} y$ if and only if $y R x$. The converse of a relation is obtained simply by reversing the order of the members of all the ordered couples which constitute the relation. Thus the converse of the relation of husband is that of wife. Referring to the simple relation R_1 introduced above,

$$\breve{R}_1 = \{\langle 1, 0 \rangle, \langle 3, 2 \rangle\}.$$

The definition is so framed that members of sets which are not ordered pairs do not belong to the converse of the set, and thus the converse of every set is a relation.

DEFINITION 6. $\breve{A} = \{\langle x, y \rangle : y A x\}$.

As usual the immediate problem is to show that the axioms stated in Chapter 2 are strong enough to guarantee existence of the intuitively appropriate converse set.

THEOREM 11. $x \breve{A} y \leftrightarrow y A x$.

PROOF.   By virtue of the axiom schema of separation

(1) $(\exists B)(\forall x)(x \in B \leftrightarrow x \in \mathcal{R}A \times \mathcal{D}A \,\&\, (\exists y)(\exists z)(x = \langle y, z \rangle \,\&\, z A y))$.

As in previous proofs the crucial step is to show that the formula:

(2) $(\exists y)(\exists z)(x = \langle y, z \rangle \,\&\, z A y)$

implies:

(3) $x \in \mathcal{R}A \times \mathcal{D}A$.

The following implications suffice, given that $x = \langle y, z \rangle$:

$$z A y \rightarrow y \in \mathcal{R}A \,\&\, z \in \mathcal{D}A$$
$$\rightarrow \langle y, z \rangle \in \mathcal{R}A \times \mathcal{D}A$$
$$\rightarrow x \in \mathcal{R}A \times \mathcal{D}A.$$

Thus we are now justified in concluding from (1) that

(4) $(\exists B)(\forall x)(x \in B \leftrightarrow (\exists y)(\exists z)(x = \langle y, z \rangle \,\&\, z A y))$.

By the routine steps previously taken (see, e.g., proof of Theorem 4), we infer from (4) and Definition 6 that

(5) $x \in \breve{A} \leftrightarrow (\exists y)(\exists z)(x = \langle y, z \rangle \ \& \ z \, A \, y)$.

Our theorem follows by a straightforward application of quantifier logic to (5).   Q.E.D.

The strategy of this proof, like others which appeal to the axiom schema of separation for establishment of the existence of some set, naturally falls into two parts. First, it must be decided what set already known to exist has the desired set as a subset. Here the answer is that the set A is intuitively a subset of the Cartesian product of the range and the domain of A. Second, it must be proved that satisfaction of the condition φ in the axiom of separation implies membership in the larger set. Here this consists of showing that (2) implies (3). When these two parts of the proof are completed, it is usually a routine matter to finish the remainder.

We now turn to some theorems about the converse operation; their intuitive content should be obvious.

THEOREM 12. $\breve{A}$ *is a relation.*

PROOF.   If $x \in \breve{A}$, then by virtue of the definition of the converse operation and Theorem 47 of Chapter 2

$$(\exists y)(\exists z) \ (x = \langle y, z \rangle);$$

the theorem then follows from the definition of relations.   Q.E.D.

THEOREM 13. $\breve{\breve{A}} \subseteq A$.

THEOREM 14. R *is a relation* $\rightarrow \breve{\breve{R}} = R$.

Three distributive laws are next.

THEOREM 15. $\overbrace{A \cap B} = \breve{A} \cap \breve{B}$.

PROOF.   By virtue of Theorem 12 and Theorem 3 it is clear that we need only consider ordered pairs:

$$x \, \overbrace{A \cap B} \, y \leftrightarrow y \, A \cap B \, x$$
$$\leftrightarrow y \, A \, x \ \& \ y \, B \, x$$
$$\leftrightarrow x \, \breve{A} \, y \ \& \ x \, \breve{B} \, y$$
$$\leftrightarrow x (\breve{A} \cap \breve{B}) y. \qquad\qquad \text{Q.E.D.}$$

(Note that in listing a sequence of equivalences we do not now justify each step when the inference is obvious from previous theorems.)

THEOREM 16. $\overbrace{A \cup B} = \breve{A} \cup \breve{B}$.

THEOREM 17. $\overbrace{A \sim B} = \breve{A} \sim \breve{B}$.

The notion which it is natural to introduce next is that of the *relative product* of two sets. If R and S are relations then the relative product of R and S (in symbols: R/S) is the relation which holds between x and y if and only if there exists a z such that R holds between x and z, and S holds between z and y. Symbolically, we have:

DEFINITION 7. $A/B = \{\langle x, y\rangle: (\exists z)(x\,A\,z\,\&\,z\,B\,y)\}.$

If, for instance,

$$R = \{\langle 1, 3\rangle, \langle 2, 3\rangle\}$$
$$S = \{\langle 3, 1\rangle\}$$

then

$$R/S = \{\langle 1, 1\rangle, \langle 2, 1\rangle\}$$
$$S/R = \{\langle 3, 3\rangle\}.$$

Proof of the next theorem, which uses the axiom schema of separation to settle the usual existence question, is left as an exercise.

THEOREM 18. $xA/By \leftrightarrow (\exists z)(x\,A\,z\,\&\,z\,B\,y).$

The proofs of the next four theorems are easy and will be left as exercises.

THEOREM 19.   *A/B is a relation.*

THEOREM 20. $0/A = 0.$

THEOREM 21. $\mathfrak{D}(A/B) \subseteq \mathfrak{D}A.$

THEOREM 22. $A \subseteq B\,\&\,C \subseteq D \rightarrow A/C \subseteq B/D.$

Three theorems asserting distributive laws follow; the proof of only one is given here.

THEOREM 23. $A/(B \cup C) = (A/B) \cup (A/C).$

PROOF.   By virtue of Theorem 19 and the previously proved fact that the union of two relations is a relation, we know at once that both $A/(B \cup C)$ and $(A/B) \cup (A/C)$ are relations. Hence the following equivalences establish our theorem:

$$\begin{aligned}
x\,A/(B \cup C)\,y &\leftrightarrow (\exists z)(x\,A\,z\,\&\,z\,B \cup C\,y)\\
&\leftrightarrow (\exists z)(x\,A\,z\,\&\,(z\,B\,y \vee z\,C\,y))\\
&\leftrightarrow (\exists z)((x\,A\,z\,\&\,z\,B\,y) \vee (x\,A\,z\,\&\,z\,C\,y))\\
&\leftrightarrow (\exists z)(x\,A\,z\,\&\,z\,B\,y) \vee (\exists z)(x\,A\,z\,\&\,z\,C\,y)\\
&\leftrightarrow x\,A/By \vee x\,A/Cy\\
&\leftrightarrow x(A/B) \cup (A/C)y. \qquad \text{Q.E.D.}
\end{aligned}$$

It may be noticed that finding a proof of this sort depends upon some familiarity with the manipulation of quantifiers.

THEOREM 24. $A/(B \cap C) \subseteq (A/B) \cap (A/C)$.

THEOREM 25. $(A/B) \sim (A/C) \subseteq A/(B \sim C)$.

The example following the definition of the relative product operation indicates that this operation is not commutative. When combined with the converse operation, we do get the following interchange of order:

THEOREM 26. $\overset{\frown}{A/B} = \breve{B}/\breve{A}$.

PROOF.

$$x \overset{\frown}{A/B} y \leftrightarrow yA/Bx$$

$$\leftrightarrow (\exists z)(y \ A \ z \ \& \ z \ B \ x)$$

$$\leftrightarrow (\exists z)(x \ \breve{B} \ z \ \& \ z \ \breve{A} \ y)$$

$$\leftrightarrow x \ \breve{B}/\breve{A} \ y. \hspace{3cm} \text{Q.E.D.}$$

The next theorem shows that the relative product operation is associative, and thus parentheses may be omitted without ambiguity in reiterated occurrences of the relative product symbol.

THEOREM 27. $(A/B)/C = A/(B/C)$.

The proof is omitted because it is a straightforward exercise in quantifier logic.

We now define the notion of a relation's domain being *restricted* to a given set. As usual the definition applies to arbitrary sets.

DEFINITION 8. $R|A = R \cap (A \times \Re(R))$.*

The definition may be illustrated by a simple example. Let

$$R = \{\langle 1, 2 \rangle, \langle 2, 3 \rangle, \langle 0, \text{Edgar Guest} \rangle\},$$

$$A = \{1, 2\}.$$

Then

$$R|A = \{\langle 1, 2 \rangle, \langle 2, 3 \rangle\}.$$

Proofs of the following six theorems are left as exercises.

THEOREM 28. $x R|A y \leftrightarrow x R y \ \& \ x \in A$.

THEOREM 29. $A \subseteq B \rightarrow R|A \subseteq R|B$.

---

*The vertical line is a standard notation for this notion. See, for instance, Kuratowski [1933, p. 12].

THEOREM 30. $R|(A \cap B) = (R|A) \cap (R|B)$.

THEOREM 31. $R|(A \cup B) = (R|A) \cup (R|B)$.

THEOREM 32. $R|(A \sim B) = (R|A) \sim (R|B)$.

THEOREM 33. $(R/S)|A = (R|A)/S$.

The next definition introduces the notation*: $R``A$, which is read: *the image of the set A under R.*   Thus if R and A are defined as in the previous example

$$R``A = \{2, 3\},$$

and

$$R``\{0\} = \{\text{Edgar Guest}\}.$$

DEFINITION 9. $R``A = \Re(R|A)$.

Most of the proofs of the theorems concerning images of sets are omitted. To reduce the number of parentheses in the statement of these theorems, we use the convention that '$\cup$', '$\cap$', '$\sim$' dominate '``'. Thus $R``A \cup B$ is $(R``A) \cup B$ not $R``(A \cup B)$.

THEOREM 34. $y \in R``A \leftrightarrow (\exists x)(x \, R \, y \, \& \, x \in A)$.

THEOREM 35. $R``(A \cup B) = R``A \cup R``B$.

PROOF. $y \in R``(A \cup B) \leftrightarrow (\exists x)(x \, R \, y \, \& \, x \in A \cup B)$

$\leftrightarrow (\exists x)(x \, R \, y \, \& \, x \in A) \vee (\exists x)(x \, R \, y \, \& \, x \in B)$

$\leftrightarrow y \in R``A \vee y \in R``B$

$\leftrightarrow y \in R``A \cup R``B.$ Q.E.D.

THEOREM 36. $R``(A \cap B) \subseteq R``A \cap R``B$.

A simple example shows that inclusion cannot be strengthened to identity in this theorem.   Let

$$R_1 = \{\langle 1, 3 \rangle, \langle 2, 3 \rangle\},$$
$$A_1 = \{1\},$$
$$B_1 = \{2\}.$$

Then

$$R_1``(A_1 \cap B_1) = 0,$$

but

$$R_1``A_1 \cap R_1``B_1 = \{3\}.$$

*The notation $R``A$ follows that of Whitehead and Russell.

THEOREM 37. $R``A \sim R``B \subseteq R``(A \sim B)$.

PROOF. $y \in R`` A \sim R``B \leftrightarrow y \in R``A \ \& \ y \notin R``B$

$$\leftrightarrow (\exists x)(x \, R \, y \ \& \ x \in A) \ \& - (\exists z)(z \, R \, y \ \& \ z \in B)$$

$$\leftrightarrow (\exists x)(x \, R \, y \ \& \ x \in A) \ \& \ (\forall z)(z \, R \, y \rightarrow z \notin B)$$

$$\rightarrow (\exists x)(x \, R \, y \ \& \ x \in A \ \& \ x \notin B)$$

$$\rightarrow (\exists x)(x \, R \, y \ \& \ x \in A \sim B)$$

$$\rightarrow y \in R``(A \sim B). \hspace{3cm} \text{Q.E.D.}$$

The particular sets used in the example just preceding this theorem may be used to show again that inclusion cannot be strengthened to identity:

$$R_1``A_1 \sim R_1``B_1 = 0$$

$$R_1``(A_1 \sim B_1) = R_1``A_1 = \{3\}.$$

THEOREM 38. $A \subseteq B \rightarrow R``A \subseteq R``B$.

THEOREM 39. $R``A = 0 \leftrightarrow \mathfrak{D}R \cap A = 0$.

The following theorem is somewhat surprising.

THEOREM 40. $\mathfrak{D}R \cap A \subseteq \breve{R}``(R``A)$.

PROOF. $x \in \mathfrak{D}R \ \cap \ A \leftrightarrow (\exists y)(x \, R \, y \ \& \ x \in A)$

$$\rightarrow (\exists y)(x \, R \, y \ \& \ y \in R``A)$$

$$\rightarrow (\exists y)(y \, \breve{R} \, x \ \& \ y \in R``A)$$

$$\rightarrow x \in \breve{R}``(R``A). \hspace{2.5cm} \text{Q.E.D.}$$

The particular sets R_1, A_1, and B_1 show why inclusion cannot be replaced by identity, and why the equivalence in line (1) of the proof must be weakened to an implication in line (2).

$$\mathfrak{D}R_1 \cap A_1 = \{1\},$$

but

$$\breve{R}_1``(R_1``A_1) = \{1, 2\}.$$

THEOREM 41. $(R``A) \cap B \subseteq R``(A \cap \breve{R}``B)$.

PROOF. $y \in (R``A) \cap B \leftrightarrow (\exists x)(x \, R \, y \ \& \ x \in A \ \& \ y \in B)$

$$\rightarrow (\exists x)(x \, R \, y \ \& \ x \in \breve{R}``B \ \& \ x \in A)$$

$$\rightarrow (\exists x)(x \, R \, y \ \& \ x \in A \cap \breve{R}``B)$$

$$\rightarrow y \in R``(A \cap \breve{R}``B). \hspace{2cm} \text{Q.E.D.}$$

Some further theorems on the restriction and image operations are given in the last section of this chapter under the additional hypothesis that R is a function.

EXERCISES

1. Give a counterexample to the statement:
$$\mathfrak{D}A = 0 \rightarrow A = 0.$$

2. Which, if any, of the analogues of Theorems 5, 6, and 7 hold for the field operation $\mathfrak{F}$?

3. Prove that the Cartesian product of two sets is a relation.

4. Prove Theorem 3.

5. Under what conditions does
$$\mathfrak{D}(A \times B) = A?$$

6. Under what conditions does
$$\mathfrak{F}(A \times B) = A \cup B?$$

7. Prove Theorems 6 and 7.

8. Give a counterexample to the statement:
$$\mathfrak{D}A \cap \mathfrak{D}B \subseteq \mathfrak{D}(A \cap B).$$

9. Prove Theorems 8, 9, and 10.

10. Give a counterexample to the statement:
$$\breve{\breve{A}} = A.$$

11. Prove that
$$\overbrace{A \times B} = B \times A.$$

12. Prove Theorems 13 and 14.

13. Prove Theorems 16 and 17.

14. Prove that
$$(A \times B)/(A \times B) \subseteq A \times B.$$

15. Prove Theorems 19-22.

16. Prove Theorems 24 and 25.

17. Give a counterexample to the statement:
$$(A/B) \cap (A/C) \subseteq A/(B \cap C).$$

18. Prove Theorem 27.

19. Prove that
$$x \in \mathfrak{D}A \rightarrow x\,A/\breve{A}x.$$

20. Is it true that

(a) $R/\cup A = \cup (R/A),$

(b) $(\forall R)(R \in A \rightarrow R/R \subseteq R) \rightarrow \cup A/\cup A \subseteq \cup A,$

(c) $(\forall R)(R \in A \to R/R \subseteq R) \to \cap A/\cap A \subseteq \cap A,$

(d) $(\forall R)(R \in A \to R/\breve{R} = R) \to \cup A/\widecheck{\cup A} = \cup A,$

(e) $(\forall R)(R \in A \to R/\breve{R} = R) \to \cap A/\widecheck{\cap A} = \cap A?$

21. Let R be the numerical relation such that

$$x \, R \, y \leftrightarrow x + y = 1.$$

Let A be the set of prime numbers between 10 and 20. Explicitly describe $R|A$.

22. Prove Theorem 28.

23. Prove Theorems 29-31.

24. Prove Theorems 32 and 33.

25. Let R be the numerical relation such that

$$x \, R \, y \leftrightarrow 2x + 1 = y.$$

Let A be the set of integers.

(a) What set is $\breve{R}``A$?

(b) What set is $R``A$?

(c) What set is $(R/R)``A$?

26. Prove Theorem 34.

27. Prove Theorem 36.

28. Prove Theorems 38 and 39.

29. Prove that $R``0 = 0$.

§ 3.2 Ordering Relations. Relations which order a set of objects occur in all domains of mathematics and in many branches of the empirical sciences. There is almost an endless number of interesting theorems about various ordering relations and their properties. Here we shall consider only certain of the more useful ones; a large number of additional theorems are included among the exercises.

We begin with the fundamental properties of reflexivity, symmetry, transitivity and the like in terms of which we define different kinds of orderings. Because these notions are so familiar, illustrative examples are given only sparingly.*

Regarding generality of definition, the situation is the same as in the last section: the definitions hold for arbitrary sets, not for just relations. However, in order to increase the immediate intuitive content of theorems, in this section we shall systematically use letters 'R', 'S', and 'T' as set variables in those contexts in which the ideas being dealt with naturally refer to relations. But it should be strictly understood that use of the variables 'R', 'S', and 'T' does not entail any formal restriction on the definitions and theorems. For instance, we define the property of transi-

*Examples and elementary applications may be found among other places in Suppes [1957, Chapter 10].

SEC. 3.2   RELATIONS AND FUNCTIONS   69

tivity for arbitrary sets R, not merely for relations. Also, without introducing a numbered definition we use henceforth the familiar notation: '$x, y \in A$' for '$x \in A$ & $y \in A$' and '$x,y,z \in A$' for '$x \in A$ & $y \in A$ & $z \in A$' etc.

We begin with eight basic definitions.

Definition 10. R *is reflexive in* $A \leftrightarrow (\forall x)(x \in A \rightarrow x \, R \, x)$.

Definition 11. R *is irreflexive in* $A \leftrightarrow (\forall x)(x \in A \rightarrow -(x \, R \, x))$.

Definition 12. R *is symmetric in* A
$$\leftrightarrow (\forall x)(\forall y)(x,y \in A \, \& \, x \, R \, y \rightarrow y \, R \, x).$$

Definition 13. R *is asymmetric in* A
$$\leftrightarrow (\forall x)(\forall y)(x,y \in A \, \& \, x \, R \, y \rightarrow -(y \, R \, x)).$$

Definition 14. R *is antisymmetric in* A
$$\leftrightarrow (\forall x)(\forall y)(x,y \in A \, \& \, x \, R \, y \, \& \, y \, R \, x \rightarrow x = y).$$

Definition 15. R *is transitive in* A
$$\leftrightarrow (\forall x)(\forall y)(\forall z)(x,y,z \in A \, \& \, x \, R \, y \, \& \, y \, R \, z \rightarrow x \, R \, z).$$

Definition 16. R *is connected in* A
$$\leftrightarrow (\forall x)(\forall y)(x,y \in A \, \& \, x \neq y \rightarrow x \, R \, y \lor y \, R \, x).$$

Definition 17. R *is strongly connected in* A
$$\leftrightarrow (\forall x)(\forall y)(x,y \in A \rightarrow x \, R \, y \lor y \, R \, x).$$

In order to relate the above eight properties to the operations introduced in the previous section, it is more elegant to consider the corresponding one-place properties, that is, to deal with relations which are reflexive, rather than reflexive in some set A, etc. The general definitions just given are useful later. For brevity we define the eight one-place properties with one fell swoop; the definitions are obvious: we simply take A to be the field of the relation.

Definition 18. R *is* $\left\{ \begin{array}{c} reflexive \\ \vdots \\ strongly \\ connected \end{array} \right\} \leftrightarrow R$ *is* $\left\{ \begin{array}{c} reflexive \\ \vdots \\ strongly \\ connected \end{array} \right\}$ *in* $\mathfrak{F}R$.

In formulating the desired theorems, we need the notion of the identity relation on a set. It is clear from Theorem 50 of Chapter 2:

$$\{x: \ x = x\} = 0$$

that we cannot define an appropriate general identity relation, but what we can do is to define, for each set A, the identity relation $\mathcal{I}A$ on A. (Thus the symbol '$\mathcal{J}$' is not an individual constant designating the identity relation but a unary operation symbol.)

DEFINITION 19. $\mathcal{J}A = \{\langle x, x\rangle: x \in A\}$.

In addition to the definition, for working purposes we need the usual theorem guaranteeing that $\mathcal{J}A$ is the empty set only when we expect it to be.

THEOREM 42. $x \,\mathcal{J}A\, x \leftrightarrow x \in A$.

PROOF. Since

$$\langle x, x\rangle = \{\{x\}, \{x, x\}\} = \{\{x\}\},$$

it is clear that

$$\mathcal{J}A \subseteq \mathcal{P}\mathcal{P}A.$$

Moreover, it is easy to show that

(1) $x \in A \rightarrow \{\{x\}\} \in \mathcal{P}\mathcal{P}A$.

By virtue of the axiom schema of separation and the definition of abstraction we may use (1) to obtain the theorem. Q.E.D.

In this and subsequent proofs which use the axiom schema of separation to prove the existence of some set, we restrict ourselves to consideration of two crucial steps: deciding what set known to exist has the desired set as a subset, and then showing that satisfaction of the appropriate condition φ in the axiom (here φ is '$x \in A$') implies membership in the larger set. It is perhaps clarifying to remark that the formal proof does not require the inference that

$$\mathcal{J}A \subseteq \mathcal{P}\mathcal{P}A,$$

although this easily follows. But the search for a set that has $\mathcal{J}A$ as a subset is an essential strategic consideration in finding a valid proof.

We state without proof three simple theorems concerning identity relations.

THEOREM 43. $\mathcal{D}\mathcal{J}A = A$.

THEOREM 44. $\mathcal{J}A/\mathcal{J}A = \mathcal{J}A$.

THEOREM 45. R is a relation $\leftrightarrow (\mathcal{J}\mathcal{D}R)/R = R$.

The eight theorems which follow could have been used as definitions, and they are often so preferred. In the proofs familiar properties of the operations are used without explicit reference to the appropriate theorems.

THEOREM 46. R is reflexive $\leftrightarrow \mathcal{J}\mathcal{F} R \subseteq R$.

PROOF. [Necessity]. By Definition 19 every element in $\mathfrak{I}\mathfrak{F}R$ is of the form $\langle x, x \rangle$, whence by Theorem 42, $x \in \mathfrak{F}R$, and it then follows from the hypothesis that R is reflexive that $\langle x, x \rangle \in R$.

[Sufficiency]. Let x be an arbitrary element of $\mathfrak{F}R$. Since our hypothesis is that

$$\mathfrak{I}\mathfrak{F}R \subseteq R$$

it follows at once that

$$\langle x, x \rangle \in R,$$

but then R is reflexive. Q.E.D.

This proof is fairly trivial, but it illustrates the approach to the remaining seven, most of which are not proved here.

THEOREM 47. *R is irreflexive* $\leftrightarrow R \cap \mathfrak{I}\mathfrak{F}R = 0$.

THEOREM 48. *R is symmetric* $\leftrightarrow \breve{\breve{R}} = \breve{R}$.

THEOREM 49. *R is asymmetric* $\leftrightarrow R \cap \breve{R} = 0$.

THEOREM 50. *R is antisymmetric* $\leftrightarrow R \cap \breve{R} \subseteq \mathfrak{I}\mathfrak{D}R$.

THEOREM 51. *R is transitive* $\leftrightarrow R/R \subseteq R$.

PROOF. [Necessity]. If

$$x\,R/R\,y$$

then there is a z such that

$$x\,R\,z\ \&\ z\,R\,y,$$

whence by the hypothesis of transitivity

$$x\,R\,y.$$

[Sufficiency]. From our hypothesis that

$$R/R \subseteq R$$

we have at once:

(1) $\qquad (\exists z)\,(x\,R\,z\ \&\ z\,R\,y) \rightarrow x\,R\,y,$

but it is a familiar fact of quantifier logic that (1) is logically equivalent to:

(2) $\qquad x\,R\,z\ \&\ z\,R\,y \rightarrow x\,R\,y.$ Q.E.D.

THEOREM 52. *R is connected* $\leftrightarrow (\mathfrak{F}R \times \mathfrak{F}R) \sim \mathfrak{I}\mathfrak{F}R \subseteq R \cup \breve{R}$.

PROOF. [Necessity]. If

(1) $\qquad x\,[(\mathfrak{F}R \times \mathfrak{F}R) \sim \mathfrak{I}\mathfrak{F}R]\,y$

then

(2) $\qquad x \in \mathfrak{F}R\ \&\ y \in \mathfrak{F}R\ \&\ x \neq y,$

but (2) together with the hypothesis that R is connected yields:

(3) $x \, R \, y \vee y \, R \, x,$

whence

(4) $x(R \cup \breve{R})y.$

[Sufficiency].   We want to derive (3) from (2) on the hypothesis that

(5) $(\mathfrak{F}R \times \mathfrak{F}R) \sim \mathfrak{g}\mathfrak{F}R \subseteq R \cup \breve{R}.$

Now (5) entails that (1) implies (4), but (1) is equivalent to (2), and (3) is equivalent to (4).   Q.E.D.

THEOREM 53.   *R is strongly connected $\leftrightarrow \mathfrak{F}R \times \mathfrak{F}R \subseteq R \cup \breve{R}$.*

Numerous additional facts are stated in the exercises: asymmetry implies irreflexivity; symmetry and transitivity imply reflexivity; all eight properties defined in Definition 18 are invariant under the converse operation; and so forth.

We now use these eight properties to define five kinds of ordering relations; the kinds are not mutually exclusive. For instance, any partial ordering is also a quasi-ordering.

DEFINITION 20.   *R is a quasi-ordering of $A \leftrightarrow R$ is reflexive and transitive in A.*

DEFINITION 21.   *R is a partial ordering of $A \leftrightarrow R$ is reflexive, antisymmetric and transitive in A.*

DEFINITION 22.   *R is a simple ordering of $A \leftrightarrow R$ is antisymmetric, transitive and strongly connected in A.*

DEFINITION 23.   *R is a strict partial ordering of $A \leftrightarrow R$ is asymmetric and transitive in A.*

DEFINITION 24.   *R is a strict simple ordering of $A \leftrightarrow R$ is asymmetric, transitive and connected in A.*

Analogous to Definition 16, we also define en masse the appropriate one-place predicates.

DEFINITION 25.

$$R \text{ is a} \left\{ \begin{array}{c} \textit{quasi-ordering} \\ \vdots \\ \textit{strict simple} \\ \textit{ordering} \end{array} \right\} \leftrightarrow R \text{ is a} \left\{ \begin{array}{c} \textit{quasi-ordering} \\ \vdots \\ \textit{strict simple} \\ \textit{ordering} \end{array} \right\} \text{of } \mathfrak{F}R.$$

Definitions 20-24 will be useful in Chapter 6, which is concerned with the construction of the real numbers. For the moment, we state some obvious theorems about the orderings of Definition 25.

Theorem 54. *R is a partial ordering $\rightarrow R$ is a quasi-ordering.*

Theorem 55. *R is a simple ordering $\rightarrow R$ is a partial ordering.*

Theorem 56. *R is a simple ordering $\rightarrow \breve{R}$ is a simple ordering.*

Theorem 57. *R and S are quasi-orderings $\rightarrow R \cap S$ is a quasi-ordering.*

proof. We need to show that $R \cap S$ is reflexive and transitive. Let x be an arbitrary element of $\mathfrak{F}(R \cap S)$. Then $x \in \mathfrak{F}R$ and $x \in \mathfrak{F}S$, whence by hypothesis

$$x\,R\,x\ \&\ x\,S\,x,$$

and thus

$$x\ R \cap S\ x.$$

Transitivity is established by the following implications, the second line following from the first on the basis of the hypothesis of the theorem:

$$x\,R \cap S\,y\ \&\ y\,R \cap S\,z \rightarrow x\,R\,y\ \&\ y\,R\,z\ \&\ x\,S\,y\ \&\ y\,S\,z$$
$$\rightarrow x\,R\,z\ \&\ x\,S\,z$$
$$\rightarrow x\,R \cap S\,z. \hspace{2cm} \text{Q.E.D.}$$

The union of any two quasi-orderings is not a quasi-ordering. For instance, let

$$R = \{\langle 1, 1\rangle, \langle 2, 2\rangle, \langle 1, 2\rangle\}$$

and

$$S = \{\langle 2, 2\rangle, \langle 3, 3\rangle, \langle 2, 3\rangle\},$$

then R and S are quasi-orderings, but $R \cup S$ is not, for it is not transitive. However, if the fields of R and S are mutually exclusive, then their union is a quasi-ordering, as asserted in the next theorem.

Theorem 58. *R and S are quasi-orderings $\&\ \mathfrak{F}R \cap \mathfrak{F}S = 0 \rightarrow R \cup S$ is a quasi-ordering.*

An exact statement of the relationship between partial orderings and strict partial orderings is provided by the next two theorems.

Theorem 59. *R is a partial ordering $\rightarrow R \sim \mathfrak{g}\mathfrak{F}R$ is a strict partial ordering.*

Theorem 60. *R is a strict partial ordering $\rightarrow R \cup \mathfrak{g}\mathfrak{F}R$ is a partial ordering.*

The sense in which a simple ordering or strict simple ordering is complete is expressed by the following theorem.

THEOREM 61. $R \subseteq S \subseteq A \times A$ & R and S are strict simple orderings of $A \to R = S$.

We now want to introduce the important notion of a relation well-ordering a set. If R is a strict simple ordering of A then R well-orders A if every non-empty subset of A has a first or minimal element (under the relation R). Actually, as we shall see, we need assume only that R is connected in A rather than it is a strict simple ordering of A. The asymmetry and transitivity of R in A are then provable, as is the fact that any element of A except the last (under the relation R) has an immediate successor.

Since this notion of well-ordering is somewhat subtler than the previous order notions introduced, the consideration of several examples will be a useful preliminary to the formal definition and theorems. In these examples, as in previous ones, we shall use integers, and in fact, real numbers, although these entities have not yet been defined formally within our system of set theory.

Let N be the set of positive integers. Then N is well-ordered by the relation *less than*, since each non-empty subset of N has a first element, namely, the smallest integer in the set. On the other hand, N is not well-ordered by *greater than*, since many subsets do not have first elements, in particular N itself. N does not have a first element with respect to $>$ just because there is no largest integer.

The notion of a well-ordering is so conceived that, unlike the other order properties considered so far, it is not invariant under the converse operation, that is, if R is a well-ordering, it does not follow that $\breve{R}$ is a well-ordering. We already have one such example: $<$ well-orders N, but $\breve{<}$ does not. A somewhat different example is given by considering

$$A = \left\{ 0, \frac{1}{2}, \frac{2}{3}, \frac{3}{4}, \cdots, \frac{n\text{-}1}{n}, \cdots, 1 \right\},$$

that is,

$$A = \left\{ \frac{n\text{-}1}{n} : n \text{ is a positive integer} \right\} \cup \{1\}.$$

The set A is well-ordered by $<$, but not by $\breve{<}$, that is, not by $>$. In this case, the set A itself has a first element under the relation $>$, but the subset $A \sim \{1\}$ does not.

By appropriate modification of the definition of R-first element of A we can frame the definition of well-orderings in such a fashion that either $<$ or $\leq$ well-orders N, that is, we can let our well-orderings be simple orderings as well as strict simple orderings. To a large extent, the choice is arbitrary; we can, if we want, let a well-ordering be neither of the two. For instance, if $A = \{1, 2, 3\}$ and

$$R = \{\langle 1, 1 \rangle, \langle 2, 2 \rangle, \langle 1, 2 \rangle, \langle 2, 3 \rangle, \langle 1, 3 \rangle\}$$

then intuitively R well-orders A even though R is neither a simple ordering nor a strict simple ordering of A. But this generalization is trivial, and there is one persuasive reason for choosing strict simple orderings rather than simple orderings: the membership relation is a strict simple ordering of the ordinal numbers as we shall define them, and, as we shall see, in Chapter 7, there is a natural connection between any well-ordering of a set and the well-ordering of the ordinals by the membership relation.

We turn now to formal developments. It is technically convenient to distinguish between the notion of a *minimal* element and that of a *first* element. A minimal element has no predecessors, whereas a first element precedes every other element. To prove asymmetry of well-orderings it is simpler to use the concept of a minimal element in their definition.

DEFINITION 26. *x is an R-minimal element of* $A \leftrightarrow x \in A$ & $(\forall y)$ $(y \in A \rightarrow -(yRx))$.

An obvious feature of this definition is that if $R \cap (A \times A)$ is empty then every member of A is an R-minimal element. However such degenerate situations are not of much interest; in the case of well-orderings, we get uniqueness of the minimal element.

DEFINITION 27. *x is an R-first element of* $A \leftrightarrow x \in A$ & $(\forall y)(y \in A$ & $x \neq y \rightarrow xRy)$.

We next define well-orderings. Subsequently we state a simple necessary and sufficient condition in terms of asymmetry and first element in place of the concept of a minimal element.

DEFINITION 28. *R well-orders* $A \leftrightarrow R$ *is connected in* A & $(\forall B)$ $(B \subseteq A$ & $B \neq 0 \rightarrow B$ *has an R-minimal element*).

We now prove that under this definition R is asymmetric and transitive.

THEOREM 62. *R well-orders* $A \rightarrow R$ *is asymmetric and transitive in* A.

PROOF. To establish asymmetry, suppose by way of contradiction that there are elements x and y in A such that $x R y$ and $y R x$. Then, contrary to the hypothesis that R well-orders A, the non-empty subset $\{x,y\}$ of A has no R-minimal element.

For transitivity, suppose for some elements $x,y,z \in A$, we have $x R y$ and $y R z$, but not $x R z$. Since R is connected in A, we must then have:

$z\,R\,x$. However, the subset $\{x,y,z\}$ does not then have an R-minimal element, for $z\,R\,x$ rules out x as the R-minimal element, $x\,R\,y$ rules out y, and $y\,R\,z$ rules out z. Whence our supposition is absurd. Q.E.D.

We leave as an exercise proof of the following three theorems.

THEOREM 63. R well-orders $A \leftrightarrow R$ is asymmetric and connected in A & $(\forall B)(B \subseteq A$ & $B \neq 0 \rightarrow B$ has an R-first element$)$.

THEOREM 64. R well-orders A & $A \neq 0 \rightarrow A$ has a unique R-first element.

THEOREM 65. R well-orders A & $B \subseteq A \rightarrow R$ well-orders B.

On the other hand it is, of course, not generally true that if R well-orders A and $S \subseteq R$, then S well-orders A.

Our next task is to prove the theorem about unique immediate successors. Two definitions are needed.

DEFINITION 29. y is an R-immediate successor of $x \leftrightarrow x\,R\,y$ & $(\forall z)$ $(x\,R\,z \rightarrow z = y \vee y\,R\,z)$.

DEFINITION 30. x is an R-last element of $A \leftrightarrow x \in A$ & $(\forall y)(y \in A$ & $x \neq y \rightarrow y\,R\,x)$.

The definition of last element is obviously similar to that of first element. In fact, we have:

THEOREM 66. x is an R-last element of $A \leftrightarrow x$ is an $\breve{R}$-first element of A.

The result concerning immediate successors can now be established.

THEOREM 67. R well-orders A & $\Im R \subseteq A$ & $x \in A$ & x is not an R-last element of $A \rightarrow x$ has a unique R-immediate successor.

PROOF.   Consider $B = \{y: x\,R\,y\}$. By hypothesis the set B is not empty, since x is not the last element of the ordering, and it is easily seen that B has a unique first element, which is the immediate successor of x. Q.E.D.

In the theory of ordinal numbers it will be convenient to have the notion of an R-section and certain facts about such sections available. The closely related notion of the R-segment of a set generated by an element is also introduced.

DEFINITION 31. B is an R-section of $A \leftrightarrow B \subseteq A$ & $A \cap \breve{R}``B \subseteq B$.

Thus a set B is an R-section of a set A if all R-predecessors in A of elements of B belong to B — obviously $\breve{R}``B$ is just the set of R-predecessors of elements of B. If, for instance,

$$A = \{1, 2, 3, 4\}$$
$$B_1 = \{1, 2\}$$
$$B_2 = 0$$
$$B_3 = \{2, 3\},$$

then B_1 and B_2 are $<$-sections of A, but B_3 is not, since $1 < 2$ and $1 \in A \sim B_3$. On the other hand, B_1 is not a $>$-section of A, since $3 > 2$ and $3 \in A \sim B_1$.

DEFINITION 32. $\; \mathcal{S}(A, R, x) = \{y: \; y \in A \; \& \; y \, R \, x\}.$

The notation: $\mathcal{S}(A, R, x)$ is read: *the R-segment of A generated by x.* The set $\mathcal{S}(A, R, x)$ is just the set of R-predecessors of x which are also members of A.

THEOREM 68. $\; x \in A \; \& \; R$ *is transitive in* $A \to \mathcal{S}(A, R, x)$ *is an R-section of A.*

PROOF. Suppose $y \in \mathcal{S}(A, R, x)$. We need to show that the R-predecessors of y which are members of A are also members of $\mathcal{S}(A, R, x)$. Let z be such an R-predecessor of y, that is,

$$z \in A \cap \breve{R} \text{``} \{y\},$$

whence

(1) $\qquad\qquad\qquad z \, R \, y.$

Since $y \in \mathcal{S}(A, R, x)$, we have:

(2) $\qquad\qquad\qquad y \, R \, x,$

and thus by the hypothesis of transitivity it follows from (1) and (2);

$$z \, R \, x,$$

from which we conclude: $z \in \mathcal{S}(A, R, x)$.   Q.E.D.

On the basis of this theorem it is easily proved that

THEOREM 69. $\; R$ *well orders* $A \to (B$ *is an R-section of* $A \; \& \; B \neq A \leftrightarrow (\exists x)(x \in A \; \& \; B = \mathcal{S}(A, R, x))).$

Some further concepts of order such as those of an *R-upper bound of x,* an *R-supremum* of x, and a *lattice* are introduced in the exercises.

EXERCISES

1. Prove the following:
   (a) R is asymmetric $\to R$ is irreflexive
   (b) R is asymmetric $\; \to R$ is antisymmetric
   (c) $\mathcal{I} A$ is symmetric and antisymmetric

(d) R is symmetric and antisymmetric relation $\rightarrow (\exists A)(R = \mathcal{I}A)$

(e) R is symmetric and transitive $\rightarrow R$ is reflexive

(f) R is strongly connected $\rightarrow R$ is connected.

2. Prove that
$$R \text{ is reflexive} \rightarrow \mathfrak{D}R = \mathfrak{D}\breve{R}.$$

3. Prove that
$$R \text{ is a relation} \rightarrow (R \text{ is symmetric} \leftrightarrow R = \breve{R}).$$

4. Prove Theorems 43–45.

5. Prove Theorems 47–49.

6. Prove Theorem 50.

7. Prove Theorem 53.

8. Prove the following:

(a) R is reflexive $\rightarrow \breve{R}$ is reflexive

(b) R and S are reflexive $\rightarrow R \cup S$ is reflexive

(c) R is irreflexive $\rightarrow \breve{R}$ is irreflexive

(d) R and S are irreflexive $\rightarrow R \cap S$, $R \cup S$, and $R \sim S$ are irreflexive

(e) R is symmetric $\rightarrow \breve{R}$ is symmetric

(f) R and S are symmetric $\rightarrow R \cap S$, $R \cup S$ and $R \sim S$ are symmetric

(g) R is asymmetric $\rightarrow \breve{R}$, $R \cap S$ and $R \sim S$ are asymmetric

(h) R is antisymmetric $\rightarrow \breve{R}$, $R \cap S$ and $R \sim S$ are antisymmetric

(i) R is transitive $\rightarrow \breve{R}$ is transitive

(j) R is connected $\rightarrow \breve{R}$ is connected

(k) R is strongly connected $\rightarrow \breve{R}$ is strongly connected

(l) $R \cup \mathcal{I} \mathfrak{F}R$ is reflexive

(m) $R \sim \mathcal{I} \mathfrak{F}R$ is irreflexive

(n) R is asymmetric $\rightarrow R \cup \mathcal{I} \mathfrak{F}R$ is antisymmetric

(o) R is antisymmetric $\rightarrow R \sim \mathcal{I} \mathfrak{F}R$ is asymmetric

(p) R is transitive $\rightarrow R \cup \mathcal{I} \mathfrak{F}R$ is transitive

(q) R is transitive and antisymmetric $\rightarrow R \sim \mathcal{I} \mathfrak{F}R$ is transitive.

9. Give a counterexample to each of the following assertions:

(a) R and S are reflexive $\rightarrow R \sim S$ is reflexive

(b) R and S are reflexive $\rightarrow R/S$ is reflexive

(c) R and S are irreflexive $\rightarrow R/S$ is irreflexive

(d) R and S are symmetric $\rightarrow R/S$ is symmetric

(e) R and S are asymmetric $\rightarrow R \cup S$ is asymmetric

(f) R and S are asymmetric $\rightarrow R/S$ is asymmetric

(g) R and S are antisymmetric $\rightarrow R \cup S$ is antisymmetric

(h) R and S are transitive $\rightarrow R \cup S$ is transitive

(i) R and S are transitive $\rightarrow R \sim S$ is transitive

(j) R and S are transitive $\rightarrow R/S$ is transitive

(k) R and S are connected $\rightarrow R \cap S$ is connected

(l) R and S are connected $\rightarrow R \cup S$ is connected

(m) R and S are connected $\rightarrow R \sim S$ is connected

(n) R and S are connected $\rightarrow R/S$ is connected.

10. Prove Theorems 54–56.

11. Prove Theorems 58–60.

12. Give a counterexample to the statement that if $R \sim \mathcal{I} \mathfrak{F}R$ is a strict partial ordering then R is a partial ordering.

13. Prove Theorem 61.

14. Consider the following sets and relations:

N = set of positive integers,
I = set of integers (negative and non-negative),
Neg = set of negative integers,
Rat = set of non-negative rational numbers,
$xR_1y \leftrightarrow x < y + 2$,
$xR_2y \leftrightarrow x < y - 2$,
$xR_3y \leftrightarrow |x| < |y| \vee (|x| = |y| \ \& \ x < y)$,
$xR_4y \leftrightarrow |x| > |y| \vee (|x| = |y| \ \& \ x > y)$.

Which of the following assertions is true? For those which are false give an explicit counterexample.

(a) $<$ well-orders Neg
(b) $>$ well-orders Neg
(c) $<$ well-orders I
(d) $<$ well-orders Rat
(e) R_1 well-orders N
(f) R_2 well-orders N
(g) R_3 well-orders I
(h) R_3 well-orders Neg
(i) $\breve{R}_3$ well-orders N
(j) $\breve{R}_4$ well-orders Neg
(k) R_4 well-orders I
(l) $\breve{R}_4$ well-orders I
(m) $\breve{R}_4$ well-orders Rat.

15. Prove Theorem 63.

16. Prove Theorems 64 and 65.

17. Consider the set

$$A = \{\langle x,y \rangle : x \text{ and } y \text{ are positive integers}\}.$$

Define a relation which well-orders A.

18. Consider the set

$$B = \{\langle x,y \rangle : x \text{ and } y \text{ are integers (negative or non-negative)}\}.$$

Define a relation which well-orders B.

19. Define a relation which well-orders the non-negative rational numbers. (Since the non-negative rational numbers are not well-ordered in magnitude, that is, well-ordered by *less than*, some other device must be used; in fact, it is essential to use the fact that every rational number is the ratio of two integers.)

20. Let

$$N = \text{set of positive integers}$$
$$S_1 = \{x : x \in N \ \& \ x < 10^6\}$$
$$xR_1y \text{ if and only if } x < y + 1.$$

Then which of the following is true?

(a) S_1 is a $<$-section of N.
(b) S_1 is a $>$-section of N.
(c) S_1 is an R_1-section of N.
(d) $\{1\}$ is an R_1-section of N.

21. Prove Theorem 69.

22. What additional ordering hypotheses if any are needed to guarantee that if A and B are R-sections of C then either $A \subseteq B$ or $B \subseteq A$?

23. Consider the following definitions:

  (i) x is an R-lower bound of $A \leftrightarrow (\forall y)(y \in A \rightarrow x \, R \, y)$.
  (ii) x is an R-infimum of $A \leftrightarrow x$ is an R-lower bound of A & $(\forall y)(y$ is an R-lower bound of $A \rightarrow y \, R \, x)$.*
  (iii) y is an R-upper bound of $A \leftrightarrow (\forall x)(x \in A \rightarrow x \, R \, y)$.
  (iv) y is an R-supremum of $A \leftrightarrow y$ is an R-upper bound of A & $(\forall x)(x$ is an R-upper bound of $A \rightarrow y \, R \, x)$.†
  (v) A is a lattice relative to $R \leftrightarrow R$ is a partial ordering of A & $(\forall x)(\forall y)(x \in A$ & $y \in A \rightarrow \{x,y\}$ has an R-supremum and an R-infimum in $A)$.

 (a) Construct two partial orderings of a set of five elements, one of which yields a lattice and one of which does not.
 (b) How many distinct lattices can be constructed out of a set of three elements?
 (c) Prove that if A is a lattice relative to R then A is a lattice relative to $\breve{R}$.
 (d) Prove that if R is a simple ordering of A then A is a lattice relative to R.
 (e) Give a counterexample to the assertion that if A is a lattice relative to R and $B \subseteq A$ then B is a lattice relative to R.
 (f) Prove that if A is a lattice relative to R_1, and B is a lattice relative to R_2, then $A \times B$ is a lattice relative to the relation R such that if $x,u \in A$ and $y,v \in B$ then

$$\langle x,y \rangle \, R \, \langle u,v \rangle \leftrightarrow x \, R_1 \, u \, \& \, y \, R_2 \, v.$$

§ 3.3 Equivalence Relations and Partitions.

A relation which is reflexive, symmetric, and transitive in a set is an *equivalence relation* on that set. The most ubiquitous example is the relation of identity. The relation of parallelism between straight lines is a familiar geometric example of an equivalence relation; the relation of congruence between figures is another. The fundamental significance of equivalence relations is that they justify the application of a general principle of abstraction: objects which are equivalent in some respect generate identical classes. Analysis of equivalence classes of objects rather than of the objects themselves is often much simpler. The family of such equivalence classes of a given set A form a *partition* of the set, i.e., is a family of mutually exclusive, non-empty subsets of A whose union equals A. Conversely, as we shall see, every partition of a set defines a unique equivalence relation on that set.

For brevity we define under the same number the appropriate one- and two-place predicates.

DEFINITION 33.

  (i) *R is an equivalence relation $\leftrightarrow R$ is a relation & R is reflexive, symmetric, and transitive;*

---

*An R-infimum of A is often called an R-greatest lower bound of A.
†An R-supremum of A is also called an R-least upper bound of A.

(ii) *R is an equivalence relation on A ↔ A = $\mathfrak{F}R$ & R is an equivalence relation.*

Unlike the ordering definitions in the last section, the present definition requires that R be a relation. The motivation for adding the additional requirement here is mainly terminological. The phrase 'R is an equivalence' is not desirable since 'equivalence' is used in several different senses in logic and set theory. On the other hand, when the phrase 'R is an equivalence relation' is used, it seems odd not to require that R be a relation. A secondary motivation is provided by the simplicity of the next theorem.

The demand in the definiens of (ii) that $A = \mathfrak{F}R$ is made for technical convenience in relating equivalence relations and partitions; the obviousness of this convenience will be apparent in the sequel.

THEOREM 70.   *R is an equivalence relation* $\leftrightarrow R/\breve{R} = R$.
The next theorem relates quasi-orderings and equivalence relations in a natural way.

THEOREM 71.   *R is a quasi-ordering* $\rightarrow R \cap \breve{R}$ *is an equivalence relation.*

The following definition introduces the notation: $R[x]$; we call $R[x]$ the *R-coset* of x. Intuitively $R[x]$ is simply the set of all objects to which x stands in the relation R. When R is an equivalence relation we also speak of $R[x]$ as the *R-equivalence class* of x.

DEFINITION 34. $R[x] = \{y : x \, R \, y\}$.

If F is the relation of fatherhood, i.e., $x \, F \, y$ if and only if x is the father of y, then

$$F[\text{George VI}] = \{\text{Elizabeth, Margaret}\}$$

and

$$F[\text{Thomas Aquinas}] = 0.$$

(Of course, F is not an equivalence relation.)
As a simple artificial example, let

$$R = \{\langle 1, 1 \rangle, \langle 2, 2 \rangle, \langle 3, 3 \rangle, \langle 1, 2 \rangle, \langle 2, 1 \rangle\}.$$

Then R is an equivalence relation and

$$R[1] = R[2] = \{1, 2\},$$

$$R[3] = \{3\}.$$

Note that in place of Definition 34, we could have used:

$$R[x] = R``\{x\}.$$

It is not customary in mathematics to be so explicit about the relation R by means of which the equivalence class $[x]$ is abstracted. However, it would be incompatible with our rules of definition to omit the free variable 'R' in the definiendum. It needs to be emphasized that the notation: $R[x]$ is non-standard and perhaps unique to this book, whereas the notation: $[x]$ is frequently used.

We have the customary theorem whose proof depends on the axiom schema of separation.

THEOREM 72. $y \in R[x] \leftrightarrow x \, R \, y$.

The following two theorems put on a systematic basis the principle of abstraction mentioned at the beginning of the section. As we shall see, these two theorems provide the essential link between equivalence relations and partitions.

THEOREM 73. $x,y \in \mathfrak{F}R \, \& \, R$ *is an equivalence relation* $\rightarrow$ $(R[x] = R[y] \leftrightarrow x \, R \, y)$.

PROOF. Assume: $R[x] = R[y]$. Since R is reflexive, we have: $y \, R \, y$, and thus by the previous theorem

$$y \in R[y],$$

whence by our assumption

$$y \in R[x];$$

and by virtue of the previous theorem again, $x \, R \, y$.

Assume now: $x \, R \, y$. Let z be an arbitrary element of $R[y]$. In view of the previous theorem we have:

$$y \, R \, z,$$

whence, since R is transitive,

$$x \, R \, z,$$

and thus

$$z \in R[x].$$

We conclude that

(1) $R[y] \subseteq R[x]$.

Now let u be an arbitrary element of $R[x]$; we have at once

$$x \, R \, u.$$

And since R is symmetric we have from our assumption:

$$y \, R \, x,$$

whence by virtue of the transitivity of R

$$y \, R \, u,$$

and

$$u \in R[y].$$

Thus

(2) $\qquad\qquad\qquad R[x] \subseteq R[y],$

and we immediately infer from (1) and (2) that

$$R[x] = R[y]. \qquad\qquad\qquad \text{Q.E.D.}$$

The above proof illustrates a strategy which is very common. We want to show that the sets $R[x]$ and $R[y]$ are identical. It is not convenient to operate with a sequence of equivalences like those used in several previous proofs. Rather our strategy is to show that any arbitrary element of $R[y]$ is a member of $R[x]$, and thus $R[y]$ is a subset of $R[x]$. Then we show that $R[x]$ is a subset of $R[y]$. These two results together establish the identity of the two sets.

The second of the two theorems mentioned shows that equivalence classes do not overlap.

THEOREM 74. *R is an equivalence relation* $\rightarrow R[x] = R[y] \lor R[x] \cap R[y] = 0$.

Note that in this theorem, unlike the preceding one, there is no need to require that x and y be in $\mathfrak{F}R$, for if $x \notin \mathfrak{F}R$ then $R[x] = 0$, and the conclusion of the theorem is satisfied.

We now turn to *partitions*. Roughly speaking, a partition of a set A is a family of mutually exclusive, non-empty subsets of A whose union equals A. For instance, if

$$A = \{1, 2, 3, 4, 5\}$$

and

$$\Pi = \{\{1, 2\}, \{3, 5\}, \{4\}\},$$

then Π is a partition of A.

Formally, we have:

DEFINITION 35. *Π is a partition of* $A \leftrightarrow \cup \Pi = A \ \& \ (\forall B)(\forall C)(B \in \Pi$ $\& \ C \in \Pi \ \& \ B \neq C \rightarrow B \cap C = 0) \ \& \ (\forall x)(x \in \Pi \rightarrow (\exists y)(y \in x)).$

The use of the letter 'Π' has no formal significance, but reflects a practice that is both customary and suggestive. Notice that the last clause of the definiens excludes both individuals and the empty set from membership in a partition. However, the empty set is a partition, namely, a partition of itself. In contrast, for non-empty sets we have:

THEOREM 75. $A \neq 0 \rightarrow \{A\}$ *is a partition of* A.

The notion of one partition being *finer* than another is often useful. The intuitive idea is that Π_1 is finer than Π_2 if every member of Π_1 is a subset of some member of Π_2 and a least one such member is a proper subset. For instance, if

$$A = \{1, 2, 3\},$$
$$\Pi_1 = \{\{1\}, \{2, 3\}\},$$
$$\Pi_2 = \{A\},$$

then Π_1 is finer than Π_2. On the other hand, if

$$\Pi_3 = \{\{1, 2\}, \{3\}\},$$

then neither Π_1 nor Π_3 is finer than the other; they are simply incomparable regarding fineness. Instead of stating that one member of Π_1 is a *proper* subset of Π_2, we may require that $\Pi_1 \neq \Pi_2$, as is done in our formal definition, which is conditional in form.

DEFINITION 36. Π_1 *and* Π_2 *are partitions of* $A \rightarrow (\Pi_1$ *is finer than* $\Pi_2 \leftrightarrow \Pi_1 \neq \Pi_2 \ \& \ (\forall A)(A \in \Pi_1 \rightarrow (\exists B)(B \in \Pi_2 \ \& \ A \subseteq B)))$.

We leave as a somewhat intriguing exercise the proof of the following theorem.

THEOREM 76. *Every set has a finest partition.*

It should be clear what is meant by 'finest partition', namely, a partition which is finer than any other partition of the set. A hint concerning the proof is to consider the power set of the given set in conjunction with the axiom schema of separation. It should be intuitively obvious what is the finest partition of any set. The problem is to prove it.

To establish in a precise manner the close connection between equivalence relations and partitions we now define a set which, when R is an equivalence relation on A, is meant to be the partition of A generated by R.

DEFINITION 37. $\Pi(R) = \{B : (\exists x)(B = R[x] \ \& \ B \neq 0\}$.

For example, if

$$A_1 = \{1, 2, 3\}$$
$$R_1 = \{\langle 1, 2 \rangle, \langle 2, 1 \rangle, \langle 1, 1 \rangle, \langle 2, 2 \rangle, \langle 3, 3 \rangle\},$$

then

$$\Pi(R_1) = \{\{1, 2\}, \{3\}\};$$

it is easily seen that R_1 is an equivalence relation on A_1, and $\mathbf{II}(R_1)$ is a partition of A_1. More generally, we have:

THEOREM 77. *R is an equivalence relation on $A \rightarrow \mathbf{II}(R)$ is a partition of A.*

We also have a theorem relating inclusion of equivalence relations and fineness of the associated partitions:

THEOREM 78. *R_1 and R_2 are equivalence relations on $A \rightarrow (R_1 \subset R_2 \leftrightarrow \mathbf{II}(R_1)$ is finer than $\mathbf{II}(R_2))$.*

Notice that if we had not required in the definition of equivalence relations that $A = \mathfrak{F}R$, then this theorem would have to be reformulated, for R_1 might contain ordered pairs whose members are not in A.

We now want to define the relation generated by a partition. The definition is general in form and is thus not restricted to partitions.

DEFINITION 38. $\mathbf{R}(\Pi) = \{\langle x, y \rangle : (\exists B)(B \in \Pi \ \& \ x \in B \ \& \ y \in B)\}$.

We have the usual theorem (which was omitted in the case of Definition 37).

THEOREM 79. $x\mathbf{R}(\Pi)y \leftrightarrow (\exists B)(B \in \Pi \ \& \ x \in B \ \& \ y \in B)$.

Corresponding to Theorem 77 we have the following:

THEOREM 80. *Π is a partition of $A \rightarrow \mathbf{R}(\Pi)$ is an equivalence relation on A.*

PROOF. First, since Π is a partition of A, given any element x of A, there is a B in Π with $x \in B$, whence $x\mathbf{R}(\Pi)x$, and thus $\mathbf{R}(\Pi)$ is reflexive in A. Second, suppose $x\mathbf{R}(\Pi)y$. Then there is a $B \in \Pi$ such that $x \in B$ and $y \in B$. Hence by Definition 38

$$y\mathbf{R}(\Pi)x,$$

and thus $\mathbf{R}(\Pi)$ is symmetric in A. Third, suppose that $x\mathbf{R}(\Pi)y$ and $y\mathbf{R}(\Pi)z$. Then there is a B such that $x \in B$ and $y \in B$, and there is a C such that $y \in C$ and $z \in C$. Since y is in both B and C, we conclude from the definition of partitions that

$$B = C,$$

and thus $z \in B$. Hence by Definition 38, $x\mathbf{R}(\Pi)z$, and we see that $\mathbf{R}(\Pi)$ is transitive in A. Q.E.D.

The next theorem shows that if we generate a partition by means of an equivalence relation R, then the equivalence relation generated by the partition is simply R again; and similarly if we begin with the equivalence relation generated by a partition, this relation generates the given partition.

THEOREM 81. *Π is a partition of $A \ \& \ R$ is an equivalence relation on $A \rightarrow (\Pi = \mathbf{II}(R) \leftrightarrow \mathbf{R}(\Pi) = R)$.*

1. Prove
    (a) $(R \cap S)[x] = R[x] \cap S[x]$
    (b) $(R \cup S)[x] = R[x] \cup S[x]$.
2. Corresponding to (a) and (b) of Exercise 1, what holds for difference of sets?
3. Prove Theorem 70.
4. Prove Theorem 71.
5. Prove Theorem 72.
6. Prove Theorem 74.
7. Let every member of A be an equivalence relation.
    (a) Is $\cap A$ an equivalence relation?
    (b) Is $\cup A$ an equivalence relation?

If so, give proof. If not, give counterexample.

8. Give two partitions of the natural numbers, one of which is finer than the other.
9. Prove Theorem 76.
10. Prove Theorem 77.
11. Prove that if R is a quasi-ordering then $\mathbf{II}(R \cap \breve{R})$ is a partition of $\mathfrak{F}R$.
12. Prove Theorem 78.
13. Prove Theorem 79.
14. Prove Theorem 81.

§ 3.4 **Functions.** Since the eighteenth century, clarification and generalization of the concept of a function have attracted much attention. Fourier's representation of "arbitrary" functions (actually piecewise continuous ones) by trigonometric series encountered much opposition; and later when Weierstrass and Riemann gave examples of continuous functions without derivatives, mathematicians refused to consider them seriously. Even today many textbooks of the differential and integral calculus do not give a mathematically satisfactory definition of functions. An exact and completely general definition is immediate within our set-theoretical framework. A function is simply a many-one relation, that is, a relation which to any element in its domain relates exactly one element in its range. (Of course, distinct elements in the domain may be related to the same element in the range.) The formal definition is obvious.

DEFINITION 39. f *is a function* $\leftrightarrow f$ *is a relation* & $(\forall x)(\forall y)(\forall z)$ $(x f y \,\&\, x f z \rightarrow y = z)$.

The use of the variable 'f' is not meant to have any formal significance. We use it here in place of 'A' or 'R' to conform to ordinary mathematical usage. To summarize our use of variables up to this point:

$$\text{'}A\text{', '}B\text{', '}C\text{', } \ldots, \text{ '}R\text{', '}S\text{', '}T\text{', } \ldots, \text{'}\Pi\text{', '}f\text{', '}g\text{', } \ldots$$

are variables (with and without subscripts) which take sets as values;

$$\text{‘}x\text{’, ‘}y\text{’, ‘}z\text{’, } \ldots$$

are variables (with and without subscripts) which take sets or individuals as values.

In the case of functions we are not content to use the notation $x\,f\,y$, but want also to have at hand the standard functional notation: $f(x) = y$, where ‘$f(x)$’ is read ‘f of x’.*

DEFINITION 40. $f(x) = y \leftrightarrow [(\text{E}!z)\,(x\,f\,z)\ \&\ x\,f\,y] \vee [-(\text{E}!z)\,(x\,f\,z)\ \&\ y = 0]$.

The definition is so framed that the notation ‘$f(x)$’ has a definite meaning for any set f and any object x. For example, if

$$f = \{\langle 1, 1\rangle, \langle 1, 2\rangle, \langle 3, 4\rangle\}$$

then

$$f(1) = 0$$
$$f(2) = 0$$
$$f(3) = 4.$$

The operation of forming the composition of two functions is so extensively used in certain branches of mathematics that various special symbols have been used for it; we use a small circle ‘$\circ$’. Thus informally,

$$(f \circ g)(x) = f(g(x)).$$

Composition is defined directly in terms of relative product; we introduce the new symbol ‘$\circ$’ rather than use the relative product symbol because the order of ‘f’ and ‘g’ in ‘$f \circ g$’ is the natural one for functions and is the reverse of that in the corresponding relative product term.

DEFINITION 41. $f \circ g = g/f$.

We have as two simple theorems:

THEOREM 82.   *f and g are functions $\rightarrow f \cap g$ and $f \circ g$ are functions.*

THEOREM 83.   *f and g are functions $\&\ x \in D(f \circ g) \rightarrow (f \circ g)(x) = f(g(x))$.*

Recalling the notion of restricting the domain of a relation, we have:

THEOREM 84. $(f \circ g)\,|A = f \circ (g|A)$.

Granted that f is a function, we may strengthen two earlier theorems on the image operation (Theorems 36 and 37).

*In mathematical logic, following the usage of Whitehead and Russell in *Principia Mathematica*, the notation: $f\text{‘}x$ is often used in place of: $f(x)$.

THEOREM 85. *f is a function* $\to \breve{f}``(A \cap B) = \breve{f}``A \cap \breve{f}``B$ & $\breve{f}``A \sim \breve{f}``B$
$= \breve{f}``(A \sim B)$.

And we may strengthen the analogue of Theorem 40 for the range of f.

THEOREM 86. *f is a function* $\to (\Re f) \cap B = f``(\breve{f}``B)$.
We also have:

THEOREM 87. *f is a function* & $A \cap B = 0 \to \breve{f}``A \cap \breve{f}``B = 0$.

We now define the notion of a 1–1 function.

DEFINITION 42. *f is 1–1* $\leftrightarrow f$ *and* $\breve{f}$ *are functions*.

We have the obvious result:

THEOREM 88. *f is 1–1* & $x_1 \in \mathfrak{D}f$ & $x_2 \in \mathfrak{D}f \to (f(x_1) = f(x_2) \leftrightarrow x_1 = x_2)$.

When f is 1–1 a simple definition of its *inverse* is possible.

DEFINITION 43. *f is 1–1* $\to f^{-1} = \breve{f}$.

Useful facts are expressed in the following five theorems.

THEOREM 89. *f is 1–1* & $x f y \to (f^{-1}(y) = x \leftrightarrow f(x) = y)$.

THEOREM 90. *f is 1–1* & $x \in \mathfrak{D}f \to f^{-1}(f(x)) = x$.

THEOREM 91. *f is 1–1* & $y \in \Re f \to f(f^{-1}(y)) = y$.

THEOREM 92. *f and g are 1–1* $\to f \cap g$ *is 1–1*.

THEOREM 93. *f and g are 1–1* & $\mathfrak{D}f \cap \mathfrak{D}g = 0$
& $\Re f \cap \Re g = 0 \to f \cup g$ *is 1–1*.

It is also desirable to define in a formal way at this point some standard mathematical language which we shall use a great deal in later chapters. We summarize it in one definition.

DEFINITION 44.

    (i) *f is a function on* (*or from*) A *to* (*or into*) $B \leftrightarrow f$ *is a function* & $\mathfrak{D}f = A$ & $\Re f \subseteq B$;

    (ii) *f is a function from* A *onto* $B \leftrightarrow f$ *is a function* & $\mathfrak{D}f = A$ & $\Re f = B$;

    (iii) *f maps* A *into* $B \leftrightarrow f$ *is a 1–1 function* & $\mathfrak{D}f = A$ & $\Re f \subseteq B$;

    (iv) *f maps* A *onto* $B \leftrightarrow f$ *is a 1–1 function* & $\mathfrak{D}f = A$ & $\Re f = B$.

The distinction between 'into' and 'onto' in this definition is standard in the mathematical literature, and has its counterpart in ordinary usage. A 1–1 function f maps A *onto* B when the range of f is the whole of B; it maps A *into* B when the range of f is only some subset of B.

We conclude this section by defining the set of all functions from B to A, which is ordinarily designated: A^B. This concept is useful in a wide variety of mathematical contexts.

DEFINITION 45. $A^B = \{f : f \text{ is a function } \& \; \mathfrak{D}f = B \; \& \; \mathfrak{R}f \subseteq A\}$.

By virtue of the axiom schema of separation we may establish the usual theorem.

THEOREM 94. $f \in A^B \leftrightarrow f \text{ is a function } \& \; \mathfrak{D}f = B \; \& \; \mathfrak{R}f \subseteq A$.

We state without proof five elementary theorems.

THEOREM 95. $A^0 = \{0\}$.

THEOREM 96. $A \neq 0 \rightarrow 0^A = 0$.

THEOREM 97. $A^B = 0 \leftrightarrow A = 0 \; \& \; B \neq 0$.

THEOREM 98. $A^{\{z\}} = \{\{\langle x, y \rangle\} : y \in A\}$.

THEOREM 99. $A \subseteq B \rightarrow A^C \subseteq B^C$.

EXERCISES

1. State and prove a necessary and sufficient condition for the union of two functions to be a function.

2. Prove Theorems 82 and 83.

3. Prove Theorem 84.

4. Prove Theorems 85 and 86.

5. Prove Theorem 87.

6. Prove Theorems 89–93.

7. Prove that if f is 1–1 then:
    (a) $f``(A \cap B) = f``A \cap f``B$,
    (b) $f``(A \sim B) = f``A \sim f``B$.

8. Given that f and g are 1–1, consider the following assertions. If an assertion is true prove it. If false, give a counterexample.
    (a) $f \cup g$ is 1–1,
    (b) $f \sim g$ is 1–1,
    (c) $f \circ g$ is 1–1,
    (d) $f \cup f^{-1}$ is 1–1,
    (e) $A \cap B = 0 \rightarrow f|A \cup g|B$ is 1–1,
    (f) $A \cap B = 0 \rightarrow f``A \cap g``B = 0$.

9. Prove Theorem 94.

10. Prove Theorems 95–99.

11. Consider Exercise 23 of §3.2, in which lattices are defined. We want to develop an equivalent formulation in terms of operations. Let A be a lattice relative to R, and $x, y \in A$. Then we define:

$$x \cap_{A,R} y = R\text{-infimum of } \{x,y\}$$

$$x \cup_{A,R} y = R\text{-supremum of } \{x,y\}.$$

Prove (where the subscripts 'A' and 'R' are dropped for brevity):

(a) $x \cap x = x$
(b) $x \cup x = x$
(c) $x \cap y = y \cap x$
(d) $x \cup y = y \cup x$
(e) $x \cap (y \cap z) = (x \cap y) \cap z$
(f) $x \cup (y \cup z) = (x \cup y) \cup z$
(g) $x \cap (x \cup y) = x$
(h) $x \cup (x \cap y) = x.$

Now to go the other way, assuming for any $x,y,z \in A$, properties (a)–(g), define:

$$x \, R' \, y \leftrightarrow x \cap y = x.$$

Prove that A is a lattice relative to R'.

12. We may formally define the lambda notation for abstraction:

   If **v** *and* **w** *are distinct variables and* **w** *does not occur in the term* **t**, *then the identity*

$$(\lambda\mathbf{v})(\mathbf{t}) = \{\langle\mathbf{v},\mathbf{w}\rangle : \mathbf{t} = \mathbf{w}\}$$

   holds.

Find the following sets:

(a) $(\lambda A)(\{x : x \in A \ \& \ A \subseteq B\})$
(b) $(\lambda A)(\{x : x \in B \sim A \ \& \ A \subseteq B\})$
(c) $(\lambda A)(\{x : x \in A \ \& \ A = 0\})$

13. Prove:

(a) $(\lambda A)(A \cap A) = 0$
(b) $(\lambda A)(A \cup A) = 0$
(c) $(\lambda A)(A \sim A) = 0$
(d) $(\lambda A)(A/A) = 0$

(The significance of (a)–(d) is that there are no sets corresponding to the set operations. For instance, in view of (a) we may not regard the operation of a set intersecting with itself as a certain set of ordered pairs. This result for the special case of sets intersecting themselves is easily generalized to show that there is no set corresponding to the binary operation of intersection for any two possibly distinct sets.)

CHAPTER 4

EQUIPOLLENCE, FINITE SETS, AND CARDINAL NUMBERS

§ **4.1 Equipollence.** The axioms listed at the end of Chapter 2 (§2.10) suffice for this section and the next, but in §4.3, on cardinal numbers, we introduce a special axiom whose use will always be indicated by a dagger '†'.

In §1.1 Cantor's notion of two sets having the same power, or, as we shall say, being *equipollent*, was mentioned as fundamental. It is fundamental because it is the basis of generalizing the notion of positive integer to that of cardinal number. Two sets are equipollent if there exists a 1–1 correspondence between them, and equipollent sets have the same cardinal number. This intuitive notion of 1–1 correspondence is easily made precise: such a correspondence is just a 1–1 function. Formally we have:*

DEFINITION 1.

(i) $A \approx B$ *under* f *if and only if* f *is a* 1–1 *function whose domain is* A *and whose range is* B;

(ii) $A \approx B$ *if and only if there is an* f *such that* $A \approx B$ *under* f.

For example, if

$$A_1 = \{1, 3, 5\}$$
$$A_2 = \{1, 7, 9\},$$

then A_1 and A_2 are equipollent. Any one of several functions will establish this:

$$f_1 = \{\langle 1, 1 \rangle, \langle 3, 7 \rangle, \langle 5, 9 \rangle\},$$

or just as well:

$$f_2 = \{\langle 1, 7 \rangle, \langle 3, 9 \rangle, \langle 5, 1 \rangle\}.$$

*In this chapter, and henceforth, we use logical symbolism only sparingly in formulating definitions and theorems, but in every case the appropriate symbolic formulation should be obvious.

It is clear that two finite sets are equipollent just when they have the same number of members. (We have, of course, not yet defined the notions of finiteness or number within our axiomatic framework.) It is also clear that if one finite set is a proper subset of another, then the two sets cannot have the same power, that is, they cannot be equipollent. However, the situation is entirely different for infinite sets. Consider, for instance, the set N of positive integers $\{1, 2, 3, \ldots\}$ and the set E of even numbers $\{2, 4, 6, \ldots\}$. Obviously, E is a proper subset of N, but E and N are equipollent, which is easily shown by considering the doubling function f such that for each positive integer n

$$f(n) = 2n.$$

We see at once that f is 1–1, $\mathfrak{D}f = N$, and $\mathfrak{R}f = E$.

The first three theorems show that equipollence has the characteristic three properties of an equivalence relation.

THEOREM 1. $A \approx A$.

PROOF. The identity function $\mathit{I}A$ is an appropriate 1–1 function. Q.E.D.

THEOREM 2.   *If $A \approx B$ then $B \approx A$.*

THEOREM 3.   *If $A \approx B$ & $B \approx C$ then $A \approx C$.*

PROOF. Let f be a 1–1 function establishing that $A \approx B$ with $\mathfrak{D}f = A$, and let g be a corresponding 1–1 function for showing that $B \approx C$ with $\mathfrak{D}g = B$. Then the function $g \circ f$ is 1–1, $\mathfrak{D}(g \circ f) = A$, and $\mathfrak{R}(g \circ f) = C$, whence $A \approx C$.   Q.E.D.

We now state a number of theorems relating equipollence to operations and relations previously introduced. These theorems make the development of cardinal arithmetic in §4.3 very simple. The first theorem is used to justify the definition of cardinal addition. The second is used to justify the definition of cardinal multiplication; the third is used to prove the commutativity of cardinal multiplication, and so forth. The order of the theorems here is nearly the same as that of the corresponding theorems for cardinal numbers in §4.3.

THEOREM 4.   *If $A \approx B$ & $C \approx D$ & $A \cap C = 0$ & $B \cap D = 0$ then $A \cup C \approx B \cup D$.*

PROOF. By hypothesis there are 1–1 functions f and g such that $A \approx B$ under f and $C \approx D$ under g. It also follows from the hypothesis that

$$\mathfrak{D}f \cap \mathfrak{D}g = 0$$

and

$$\mathfrak{R}f \cap \mathfrak{R}g = 0,$$

whence by virtue of Theorem 93 of §3.4, $f \cup g$ is 1–1, and it is easily seen that

$$A \cup C \approx B \cup D \text{ under } f \cup g. \qquad \text{Q.E.D.}$$

THEOREM 5. *If $A \approx B$ & $C \approx D$ then $A \times C \approx B \times D$.*

PROOF. Let $A \approx B$ under the function f and $C \approx D$ under the function g. Then the function h such that for $x \in A$ and $y \in C$

$$h(\langle x, y \rangle) = \langle f(x), g(y) \rangle$$

establishes the equipollence of $A \times C$ and $B \times D$.   Q.E.D.

THEOREM 6. $A \times B \approx B \times A$.

PROOF. The function f such that for $x \in A$ and $y \in B$

$$f(\langle x, y \rangle) = \langle y, x \rangle$$

is appropriate to establish the desired equipollence.   Q.E.D.

THEOREM 7. $A \times (B \times C) \approx (A \times B) \times C$.

THEOREM 8. $A \times \{x\} \approx A$ & $\{x\} \times A \approx A$.

PROOF. For the first half of the theorem the function f defined on $A \times \{x\}$ such that for $y \in A$

$$f(\langle y, x \rangle) = y$$

is appropriate. A similar function is suitable for the second half.   Q.E.D.

THEOREM 9.   *There are sets C and D such that $A \approx C$ & $B \approx D$ & $C \cap D = 0$.*

PROOF. Define

$$C = A \times \{0\}$$
$$D = B \times \{\{0\}\}.$$

Then by the preceding theorem $A \approx C$ and $B \approx D$, and clearly $C \cap D = 0$. Q.E.D.

The next theorem is used to justify the definition of cardinal exponentiation.

THEOREM 10.   *If $A \approx B$ & $C \approx D$ then $A^C \approx B^D$.*

PROOF. By hypothesis there are 1–1 functions f and g such that $A \approx B$ under f and $C \approx D$ under g. If $h \in A^C$ then

(1) $$f \circ h \in B^C$$

and from (1) we infer

$$f \circ h \circ g^{-1} \in B^D.$$

Moreover, if $h' \in B^D$ then there is a unique $h \in A^C$ such that $h' = f \circ h \circ g^{-1}$ (in fact, $h = f^{-1} \circ h' \circ g$). Whence if we define the function f' on A^C such that for every $h \in A^C$

$$f'(h) = f \circ h \circ g^{-1}$$

then f' is 1–1 and its range is B^D. Q.E.D.

We omit proof of the following three theorems, which correspond to three fundamental laws of cardinal exponentiation.

THEOREM 11. *If* $B \cap C = 0$ *then* $A^{B \cup C} \approx A^B \times A^C$.

THEOREM 12. $(A \times B)^C \approx A^C \times B^C$.

THEOREM 13. $(A^B)^C \approx A^{B \times C}$.

We now anticipate the definition of the integer 2 given in Chapter 5 in order to formulate a classical theorem in standard notation: the power set $\mathcal{P}A$ is equipollent to 2^A.

THEOREM 14. *If* $2 = \{0, \{0\}\}$ *then* $\mathcal{P}A \approx 2^A$.

PROOF. Let $B \in \mathcal{P}A$. Then there is a function $g_B \in 2^A$ such that

$$g_B(x) = \begin{cases} 0 \text{ if } x \in B \\ \{0\} \text{ if } x \in A \sim B. \end{cases}$$

It is easily seen that to each B there corresponds a unique g_B, and for each $h \in 2^A$ there is a unique $B \in \mathcal{P}A$ such that $h = g_B$, which establishes the desired correspondence. Q.E.D.

We now define in the obvious manner the relation $\preceq$ of *being equal to or less than in power*, which we may also call *being equal to or less pollent than* although the phrase 'less pollent' is not standard.

DEFINITION 2. $A \preceq B$ *if and only if there is a set* C *such that* $A \approx C$ *&* $C \subseteq B$.

Three simple theorems are:

THEOREM 15. *If* $A \approx B$ *then* $A \preceq B$.

THEOREM 16. *If* $A \subseteq B$ *then* $A \preceq B$.

THEOREM 17. *If* $A \preceq B$ *&* $B \preceq C$ *then* $A \preceq C$.

A less obvious, but fundamental theorem for the Cantor theory of power is the following, whose proof is the most difficult of any theorem yet stated in these first four chapters.

THEOREM 18. [Schröder-Bernstein Theorem]* *If $A \preceq B$ & $B \preceq A$ then $A \approx B$.*

PROOF. Following the hypothesis of the theorem let

$$f \text{ map } A \text{ onto } B_1 \subseteq B,$$

and

$$g \text{ map } B \text{ onto } A_1 \subseteq A.$$

We can show that A and B have the same power if we can find a subset K of A such that g maps $B \sim f``K$ onto $A \sim K$. For the h defined as follows will then yield the appropriate correspondence

$$h = (f|K) \cup (\breve{g}|(A \sim K)),$$

since

$$\text{Domain of } h = K \cup (A \sim K) = A,$$

and

$$\text{Range of } h = (f``K) \cup (\breve{g}``(A \sim K))$$
$$= (f``K) \cup (B \sim f``K) = B.$$

In other words, we need to find a subset K of A such that

$$g``(B \sim f``K) = A \sim K.$$

To this end we now show that if we define

$$D = \{C: \ C \subseteq A \ \& \ g``(B \sim f``C) \subseteq A \sim C\},$$

then $\cup D$ is an appropriate K.

We first observe that if $C_1 \subseteq A$ & $C_2 \subseteq A$ and $C_1 \subseteq C_2$ then

$$g``(B \sim f``C_2) \subseteq g``(B \sim f``C_1),$$

whence

(1) $$A \sim g``(B \sim f``C_1) \subseteq A \sim g``(B \sim f``C_2).$$

Moreover for the special case of $C \in D$, we have

(2) $$C \subseteq A \sim g``(B \sim f``C).$$

(This follows from the definition of D and the fact that for any subsets X and Y of A, $X \subseteq A \sim Y$ if and only if $Y \subseteq A \sim X$.)

---

*The theorem was proved independently by E. Schröder and F. Bernstein in the 1890's. Because the theorem was conjectured by Cantor, it is sometimes called the Cantor-Bernstein theorem. The proof here follows the one given by Fraenkel [1953, pp. 102-3], which is credited by Fraenkel to J. M. Whitaker.

Since every $C \in D$ is a subset of $\cup D$, we conclude from (1) and (2) if $C \in D$ then

(3) $$C \subseteq A \sim g''(B \sim f'' \cup D).$$

By Theorem 63 of Chapter 2 we may infer from (3) that

(4) $$\cup D \subseteq A \sim g''(B \sim f'' \cup D).$$

Now let

(5) $$F = A \sim g''(B \sim f'' \cup D).$$

Then by virtue of (1), (4), and (5)

$$A \sim g''(B \sim f'' \cup D) \subseteq A \sim g''(B \sim f''F),$$

that is,

$$F \subseteq A \sim g''(B \sim f''F),$$

whence we conclude

$$F \in D,$$

that is

(6) $$A \sim g''(B \sim f'' \cup D) \subseteq \cup D.$$

From (4) and (6), we have

$$\cup D = A \sim g''(B \sim f'' \cup D),$$

which for $K = \cup D$ is equivalent to:

$$g''(B \sim f''K) = A \sim K,$$

the desired conclusion. Q.E.D.

We shall have occasion to use the Schröder-Bernstein theorem in proofs of several subsequent theorems. For the moment we complete our list of theorems on the relation $\leq$. Mainly, we have the following monotonicity result:

THEOREM 19. *If $A \leq B$ & $C \leq D$ then*

(i)   *if $B \cap D = 0$ then $A \cup C \leq B \cup D$,*
(ii) $A \times C \leq B \times D$,
(iii) $A^C \leq B^D$, *provided it is not the case that $A = B = C = 0$ & $D \neq 0$.*

As an immediate consequence of (i) of this theorem, we obtain:

THEOREM 20. $A \leq A \cup B$.

We now define in the expected manner the relation $<$ of *having less power*. We say 'not $B \leq A$' as an abbreviation for 'it is not the case that $B \leq A$'.

DEFINITION 3.   *A < B if and only if A $\leq$ B & not B $\leq$ A.*

Three simple results are summarized in the next theorem.

THEOREM 21.

   (i)   *Not A < A;*
  (ii)   *If A < B then not B < A;*
 (iii)   *If A < B & B < C then A < C.*

Some relations between equipollence and relative power are summarized in the following.

THEOREM 22.

   (i)   *If A $\leq$ B then not B < A;*
  (ii)   *If A $\leq$ B & B < C then A < C;*
 (iii)   *If A < B & B $\leq$ C then A < C;*
 (iv)   *A $\leq$ B if and only if either A $\simeq$ B or A < B.*

PROOF. We prove only (iv). [Necessity]. Suppose not $A < B$.

Then in view of Definition 3, either not $A \leq B$ or $B \leq A$, but by hypothesis $A \leq B$ and thus also $B \leq A$. From the Schröder-Bernstein theorem we conclude $A \simeq B$.

[Sufficiency]. If $A \simeq B$ then obviously $A \leq B$. If $A < B$ then by Definition 3, $A \leq B$ immediately. Q.E.D.

Of fundamental importance is Cantor's theorem which asserts that every set has less power than its power set.

THEOREM 23. $A < \mathcal{P}A.$

PROOF. The proof makes use of the basic argument employed in constructing Russell's paradox in intuitive set theory, although Cantor's proof was historically earlier than Russell's paradox.

The function f on A such that for x in A

$$f(x) = \{x\}$$

establishes that

(1) $A \leq \mathcal{P}A,$

since the set of unit sets of elements of A is a subset of the power set of A.

Now suppose $A \simeq \mathcal{P}A$ under the function g, say ($\mathfrak{D}g = A$ & $\mathfrak{R}g = \mathcal{P}A$). Define:

$$B = \{y : y \in A \ \& \ y \notin g(y)\}.$$

Hence on the basis of our supposition there must be an x in A such that

$$g(x) = B.$$

But we then easily infer that

$$x \in g(x) \text{ if and only if } x \notin g(x),$$

which is absurd. We conclude that our supposition is false and that it is not the case that $A \approx \mathcal{P}A$; from this result, (1) and the contrapositive of the Schroder-Bernstein theorem that it is not the case $\mathcal{P}A \preceq A$, we infer the theorem. Q.E.D.

The most important fact about the relative power of sets which we have not yet established is that the relative power of two sets is always comparable, that is, we always have: $A \prec B, A \approx B$, or $B \prec A$. This result, known as the law of trichotomy, not only requires the axiom of choice in its proof, but is equivalent to it. We defer its proof until Chapter 8 where we discuss the axiom of choice in more detail. The importance of the trichotomy for the classical theory of cardinal numbers should be obvious: without it any two cardinals are not necessarily comparable, a state of affairs which is unsatisfactory in any theory of quantity, finite or infinite.

EXERCISES

1. Prove Theorem 2.
2. Prove Theorem 7.
3. Prove Theorems 11-13.
4. Prove that if f is a function then $\mathcal{D}f \approx f$.
5. Prove Theorems 15-17.
6. Prove that if $B \neq 0$ then $A \preceq A \times B$.
7. Use the Schröder-Bernstein theorem to prove that if $A \subseteq B \subseteq C$ & $A \approx C$ then $B \approx C$. Conversely, show that this result implies the Schröder-Bernstein theorem.
8. Prove Theorem 19.
9. Prove Theorem 21.
10. Prove Parts (i)–(iii) of Theorem 22.
11. Prove that if $B \approx C$ & $A \prec B$ then $A \prec C$.
12. Prove that if $x \notin A$ & $y \notin B$ & $A \cup \{x\} \approx B \cup \{y\}$ then $A \approx B$.

§ **4.2 Finite Sets.** The common sense notion is that a set is finite just when it has exactly n members for some non-negative integer n. If it is not finite, then it is infinite. This common sense idea is technically sound and we shall use it subsequently, but it is also instructive to attempt to find a definition of finitude which does not require explicit reference to numbers. Such an approach is also intuitively sound since we continually judge that sets are finite without any clear idea of their cardinality. For instance, everyone believes that the set of hairs in the heads of all blue-eyed, left-handed Anglican clergymen in 1900 is finite, but probably no one could estimate with any accuracy the cardinality of this set.

Dedekind [1888] proposed such a non-numerical definition.* A finite set is one which is not equipollent to any of its proper subsets. Consideration of simple examples of finite sets suggests that this definition is intuitively sound. Notice, of course, that we are not looking for some arbitrary definition of finiteness, but for a definition which makes exactly those sets finite which are finite in the sense of having n members for some integer n.

As sound as Dedekind's definition may seem, it requires the axiom of choice to prove that every Dedekind finite set is finite in the ordinary sense. This proof will be given in Chapter 8.

Several other alternative non-numerical definitions of finitude have been proposed by Zermelo, Russell, Sierpinski, Kuratowski, and Tarski, to mention the more well-known ones. A complete and systematic survey is given in Tarski [1924b], where Tarski's own definition is proposed. Since Tarski's definition is simple and does not require the axiom of choice to prove its equivalence to the ordinary numerical definition, we shall adopt it here. The developments in this section closely follow Tarski's article.

The idea is that a set is finite when any non-empty family of subsets of the given set has a member of which no other member of the family is a proper subset. That is, in the terminology of Definition 26 of Chapter 3, every non-empty family of subsets has a minimal element with respect to the relation $\subset$ of being a proper subset. Formally, we define both minimal and maximal elements.

DEFINITION 4

   (i)   *x is a minimal element of A if and only if $x \in A$ & x is a set & for every B, if $B \in A$ then not $B \subset x$;*

   (ii) *x is a maximal element of A if and only if $x \in A$ & x is a set & for every B, if $B \in A$ then not $x \subset B$.*

For instance, if

$$A = \{1, 2, 3\},$$
$$K_1 = \{\{1, 2\}, \{1\}, \{3\}\}$$

and

$$K_2 = \{0, \{1, 3\}, A\},$$

then the sets $\{1\}$ and $\{3\}$ are minimal elements of K_1, and the empty set is the minimal element of K_2. Obviously any other non-empty family of subsets of A has a minimal element. On the other hand, consider the

*This approach was suggested independently by Peirce at approximately the same time. (See Peirce [1932, Vol. III, pp. 210-249].)

set N of positive integers, and consider the family F of subsets $\{N_1, N_2, \ldots, N_n, \ldots\}$ where N_n is

$$N \sim \{1, 2, \ldots, n\text{-}1\}.$$

Then clearly F has no minimal element, and this situation is typical of infinite sets. The maximal elements of K_1 are the sets $\{1, 2\}$ and $\{3\}$, and the unique maximal element of K_2 is the set A itself.

Tarski's definition, unlike Dedekind's, does not require the notion of equipollence. However, later in this section we shall prove some theorems about equipollence and relative power of finite sets which will bring us back to the ideas of the preceding section.

DEFINITION 5. *A is finite if and only if every non-empty family of subsets of A has a minimal element.*

We turn now to some elementary theorems. Proofs of the first two are quite simple.

THEOREM 24. *The empty set is finite.*

THEOREM 25. $\{x\}$ *is finite.*

THEOREM 26. *If A is finite and $B \subseteq A$ then B is finite.*

PROOF. Let K be a non-empty family of subsets of B. Since $B \subseteq A$, K is a non-empty family of subsets of A and by the hypothesis of the theorem must have a minimal element. Q.E.D.

THEOREM 27. *If A is finite then $A \cap B$ and $A \sim B$ are finite.*

PROOF. We note that

$$A \cap B \subseteq A,$$
$$A \sim B \subseteq A,$$

and then apply the preceding theorem. Q.E.D.

The proof that the union of two sets is finite is more difficult.

THEOREM 28. *If A and B are finite then $A \cup B$ is finite.*

PROOF. Let K be any non-empty family of subsets of $A \cup B$. To establish the theorem it is necessary to show that K has a minimal element. We define:

(1) $L = \{C: C \subseteq A \ \& \ \exists D \subseteq B$ such that $C \cup D \in K\}$.

The following consideration shows that L is non-empty. Let E be some element of K. Then $C = E \cap A$ is a member of L, for we may take D to be $E \sim A$.

Because A is finite, L has a minimal element, say C^*. We note:

$$(2) \qquad\qquad C^* \in L$$

$$(3) \qquad\qquad C^* \subseteq A.$$

We now define

$$M = \{E: \; E \subseteq B \, \& \, E \cup C^* \in K\}.$$

By virtue of (2) and (3), M is non-empty, and because B is finite, M, like L, has a minimal element, say E^*. We now have:

$$(4) \qquad\qquad E^* \in M$$

$$(5) \qquad\qquad E^* \subseteq B$$

$$(6) \qquad\qquad E^* \cup C^* \in K.$$

To complete the proof we show that $E^* \cup C^*$ is a minimal element of K. Suppose, by way of contradiction that there is a set G such that

$$(7) \qquad\qquad G \in K$$

$$(8) \qquad\qquad G \subset E^* \cup C^*.$$

Now from (3) and (5) we see that

$$(9) \qquad\qquad G \cap C^* \subseteq C^* \subseteq A$$

$$(10) \qquad\qquad G \cap E^* \subseteq E^* \subseteq B$$

and from (8)

$$(11) \qquad\qquad G = (G \cap C^*) \cup (G \cap E^*).$$

From (7), (9), (11) and the definition of L we infer

$$G \cap C^* \in L,$$

and since C^* is a minimal element of L

$$(12) \qquad\qquad G \cap C^* = C^*.$$

Now from (7), (11) and (12)

$$(G \cap E^*) \cup C^* \in K,$$

and thus, recalling (10),

$$G \cap E^* \in M.$$

Because E^* is a minimal element of M,

$$(13) \qquad\qquad G \cap E^* = E^*.$$

From (11), (12) and (13) we conclude that

$$G = C^* \cup E^*,$$

which contradicts (8) and proves our supposition false. Q.E.D.

As an immediate consequence of Theorems 25 and 28, we have:

THEOREM 29. *If A is finite then* $A \cup \{x\}$ *is finite.*

We turn now to the formulation and proof of a principle of induction for finite sets. As we shall see, the principle may be used as a definition of finite sets, a fact which was first recognized by Whitehead and Russell in *Principia Mathematica* [Vol. II, *120.23]. A schematic formulation of the principle of induction for the non-negative integers goes as follows:
If

   (i) $\varphi(0)$
   (ii) $(\forall n)(\varphi(n) \to \varphi(n + 1))$
then $(\forall n)\varphi(n)$.

In the next chapter we prove this principle for the integers. The corresponding principle of induction for a finite set adds the hypothesis that the set is finite and replaces (ii) by

$$(\forall x)(\forall B)(x \in A \ \& \ B \subseteq A \ \& \ \varphi(B) \to \varphi(B \cup \{x\})),$$

the idea being that if the set A is finite we start with the empty set and add elements of A, one at a time, until we have exhausted A.

The proof of the theorem about induction for finite sets is facilitated by having available the already defined notion of *maximal element* of a family of subsets. Proofs of the two theorems about maximal elements we leave as exercises.

THEOREM 30. *Every non-empty family of subsets of a finite set has a maximal element.*

The second theorem is the converse of Theorem 30 and the two together thus show that a finite set may be defined in terms of every non-empty family of its subsets having a maximal element.

THEOREM 31. *If every non-empty family of subsets of a set A has a maximal element then A is finite.*

We are now ready for the first induction theorem.

THEOREM SCHEMA 32. *If*

   (i)   *A is finite,*
   (ii) $\varphi(0)$,
   (iii) $(\forall x)(\forall B)(x \in A \ \& \ B \subseteq A \ \& \ \varphi(B) \to \varphi(B \cup \{x\}))$
then $\varphi(A)$.

PROOF. Suppose (i) and (ii) hold. We define

(1) $K = \{B : \ B \subseteq A \ \& \ \varphi(B)\}$.

The set K is not empty since $0 \subseteq A$ and $\varphi(0)$, and thus $0 \in K$. Whence by virtue of Theorem 30 and (i), K has a maximal element, say B. We want to show that $B = A$, and thus $\varphi(A)$. Suppose it were the case that $B \neq A$. On the basis of (1), $B \subseteq A$ and thus

$$A \sim B \neq 0.$$

Let $x \in A \sim B$. Then $B \cup \{x\} \subseteq A$ and by virtue of (iii) $\varphi(B \cup \{x\})$, whence

$$B \cup \{x\} \in K,$$

which contradicts B being a maximal element of K and proves our supposition false.   Q.E.D.

By taking $\varphi(B)$ as '$B \in K$', we immediately infer from Theorem 32 the following "set" formulation of induction for finite sets. It is worth noting that the converse holds; namely, Theorem 32 follows from Theorem 33 if we set $K = \{B : B \in \mathcal{P}A \,\&\, \varphi(B)\}$.

Theorem 33.   *If*

    (i)   *A is finite,*
    (ii) $0 \in K$,
    (iii) $(\forall x)(\forall B)(x \in A \,\&\, B \subseteq A \,\&\, B \in K \to B \cup \{x\} \in K)$
then $A \in K$.

That this inductive property of finite sets may be used to characterize them is established by the next theorem.

Theorem 34.   *A is finite if and only if A belongs to every set K satisfying* (ii) *and* (iii) *of Theorem* 33.

proof. The necessity follows from Theorem 33. To prove sufficiency, assume that A belongs to every set K satisfying (ii) and (iii). Let K_1 be the family of all finite subsets of A. By virtue of Theorem 24, $0 \in K_1$. Moreover if $B \in K_1$ and $x \in A$ then by Theorem 29 $B \cup \{x\} \in K_1$, whence by hypothesis $A \in K_1$ and is thus finite.   Q.E.D.

The definition of Sierpinski [1918], as modified slightly by Tarski, is given in the following theorem whose proof is similar to the preceding one.

Theorem 35.   *A is finite if and only if A belongs to every set K such that*

    (i) $0 \in K$,
    (ii) *if $x \in A$, then $\{x\} \in K$,*
    (iii) *if $B \in K \,\&\, C \in K$ then $B \cup C \in K$.*

The definition of Sierpinski is modified in Kuratowski [1920] to yield the following theorem, whose proof we leave as an exercise.

THEOREM 36. *A is finite if and only if the power set $\mathcal{P}A$ is the only set K which satisfies the conditions:*

(i) $K \subseteq \mathcal{P}A,$

(ii) $0 \in K,$

(iii)   *if* $x \in A$ *then* $\{x\} \in K,$

(iv)   *if* $B \in K \ \& \ C \in K$ *then* $B \cup C \in K.$

Our next objective is to prove some facts concerning finiteness about the power set and sum set of a given set. We first prove a preliminary theorem asserting that if we can map A onto B and A is finite then B is finite.

THEOREM 37. *If A is finite and f is a function whose domain is A and range is B then B is finite.*

PROOF. We define:

$$K = \{C: \ C \subseteq A \ \& \ f``C \text{ is finite}\}.$$

We want to prove by induction (using Theorem 33) that $A \in K$, whence B is finite, since $f``A = B$. First we observe immediately that $0 \in K$ because $0 \subseteq A$ and $f``0 = 0$. Assume, for the second part of the induction, that $x \in A$ and $C \in K$. We need to show that $C \cup \{x\} \in K$. Obviously, $C \cup \{x\} \subseteq A$. Since f is a function, $f``\{x\}$ is a unit set and thus finite (Theorem 25), and since $C \in K$, $f``C$ is finite. Therefore, in view of Theorem 28, $(f``C) \cup f``\{x\}$ is finite. But by virtue of Theorem 35 of Chapter 3

$$f``(C \cup \{x\}) = (f``C) \cup f``\{x\},$$

and $f``(C \cup \{x\})$ is finite. We conclude that

$$C \cup \{x\} \in K,$$

which completes the proof. Q.E.D.

THEOREM 38. *If a set is finite then its power set is finite.*

PROOF. As in the preceding proof, we define a certain set and then prove by induction that A is a member of it. We define:

$$K = \{B: \ B \subseteq A \ \& \ \mathcal{P}B \text{ is finite}\}.$$

Again it is obvious that $0 \in K$. Assume, as usual, for the second part of the induction that $B \in K$ and $x \in A$. If $x \in B$ the induction is immediate, so we may suppose that $x \notin B$. Clearly, $B \cup \{x\} \subseteq A$. It remains to show that the power set $\mathcal{P}(B \cup \{x\})$ is finite, given that $\mathcal{P}B$ is finite, in order to prove that $B \cup \{x\} \in K$. Let us define:

(1) $f = \{\langle C, C \cup \{x\}\rangle : \ C \in \mathcal{P}B\}.$

We show that f is a function whose range is $\mathcal{P}(B \cup \{x\}) \sim \mathcal{P}B$. (Obviously the domain of f is $\mathcal{P}B$.) That f is a function is evident from (1). The problem is to show that its range is the desired set. If $C \in \mathcal{P}B$, then

$$C \cup \{x\} \in \mathcal{P}(B \cup \{x\}) \sim \mathcal{P}B$$

since $x \notin B$. On the other hand, if

$$D \in \mathcal{P}(B \cup \{x\}) \sim \mathcal{P}B$$

then

$$D \sim \{x\} \in \mathcal{P}B,$$

whence

$$\langle D \sim \{x\}, D \rangle \in f.$$

We conclude that $\mathcal{P}(B \cup \{x\}) \sim \mathcal{P}B$ is the range of f. Applying now the preceding theorem and using the inductive hypothesis that $\mathcal{P}B$ is finite, we infer that $\mathcal{P}(B \cup \{x\}) \sim \mathcal{P}B$ is finite.

We note that

$$\mathcal{P}(B \cup \{x\}) = (\mathcal{P}(B \cup \{x\}) \sim \mathcal{P}B) \cup \mathcal{P}B$$

and since by virtue of Theorem 28 the union of two finite sets is finite, we obtain that $\mathcal{P}(B \cup \{x\})$ is finite, whence

$$B \cup \{x\} \in K$$

and the induction is completed. Q.E.D.

We leave as an exercise the inductive proof of the following theorem.

THEOREM 39. *If A is finite and every set which is a member of A is finite, then $\bigcup A$ is finite.*

The last two theorems may be used to prove each other's converse, that is, Theorem 39 is useful in proving the converse of Theorem 38, and vice versa. We leave these two converses as exercises.

THEOREM 40. *If $\mathcal{P}A$ is finite then A is finite.*

THEOREM 41. *If A is a family of sets and $\bigcup A$ is finite then A is finite and every set which is a member of A is finite.*

Notice that Theorem 41 is not the exact converse of Theorem 39, for the additional hypothesis is required that A be a family of sets. Obviously $\bigcup A$ could be finite and A infinite as long as A has only a finite number of sets as members, since individuals in A do not contribute to $\bigcup A$.

We now consider some theorems about equipollence of finite sets. The proof of the first one follows at once from Theorem 37.

THEOREM 42.   *If A is finite and $A \approx B$ then B is finite.*

THEOREM 43.   *If A is finite and $B \leq A$ then B is finite.*

PROOF.   Given that $B \leq A$, then there is a subset C of A such that $B \approx C$; but since A is finite, in view of Theorem 26, C is finite, and thus by the preceding theorem and symmetry of equipollence, B is finite.   Q.E.D.

It was mentioned at the end of the last section that the law of trichotomy, that is, the theorem that for any two sets A and B we always have: $A \prec B$, $A \approx B$ or $B \prec A$, is equivalent to the axiom of choice. However, without the axiom of choice we may prove by induction the law of trichotomy if one of the sets is finite.

THEOREM 44.   *If A is finite then*

$$A \prec B, \; A \approx B \; or \; B \prec A.$$

PROOF.   The induction is on subsets of A.   Let us define:

$$K = \{C \colon C \subseteq A \; \& \; (C \prec B, \, C \approx B \; or \; B \prec C)\}.$$

Clearly $0 \in K$, for if $B = 0$ then $0 \approx B$ and if $B \neq 0$ then $0 \prec B$. For the second part of the induction, we assume as usual that $C \in K$ and $x \in A$, and we want to show that $C \cup \{x\} \in K$. The non-trivial case is when $x \notin C$. By hypothesis

$$C \prec B, \quad C \approx B, \; or \; B \prec C.$$

These three possibilities lead to three cases.

Case 1. $C \prec B$.   Then there is a proper subset D of B such that $C \approx D$, under function f, say. Since D is a proper subset of B, there is a y in $B \sim D$, whence

$$f \cup \{\langle x, y \rangle\}$$

establishes that

$$C \cup \{x\} \approx D \cup \{y\}$$

and thus

$$C \cup \{x\} \leq B,$$

which implies:

$$C \cup \{x\} \approx B \; or \; C \cup \{x\} \prec B,$$

and therefore

$$C \cup \{x\} \in K.$$

Case 2. $C \approx B$.   Because $C \subseteq C \cup \{x\}$, we have either $C \approx C \cup \{x\}$ or $C \prec C \cup \{x\}$. If the first alternative holds then $C \cup \{x\} \approx B$. If the second, then $B \prec C \cup \{x\}$. For either one it follows that $C \cup \{x\} \in K$.

Case 3. $B < C$. Since $C \leq C \cup \{x\}$, we have at once that $B < C \cup \{x\}$, and thus $C \cup \{x\} \in K$.   Q.E.D.

It is an easy consequence of the theorem just proved (and some preceding results) that a finite set always has less power than a set which is not finite.

THEOREM 45.   *If A is finite and B is not, then $A < B$.*

We may now prove the important theorem that a finite set (in the sense of Tarski) is a Dedekind finite set. As already remarked, every known proof of the converse of this theorem requires the axiom of choice. For future reference it is convenient to define formally Dedekind finiteness.

DEFINITION 6.   *A set is Dedekind finite if and only if it is not equipollent to any of its proper subsets.*

And we prove:

THEOREM 46.   *If a set is finite then it is Dedekind finite.*

PROOF. Suppose, contrary to the theorem, that A is a finite set with a proper subset B such that

$$A \approx B$$

under the function f, say. We define

$$K = \{C: C \subseteq A \,\&\, f``C \subset C\}.$$

Because $f``A = B$ and $B \subset A$, we see at once that $A \in K$. Thus the family K of subsets of A is non-empty, and since A is finite, K has a minimal element, say D. Consequently

(1) $f``D \notin K$

for $f``D \subset D$ and $D$ is a minimal element. On the other hand, in view of the fact that $f$ is 1–1 and $f``D \subset D$, we have that

$$f``(f``D) \subset f``D,$$

and thus

(2) $f``D \in K$

but (1) and (2) are jointly absurd, and our supposition is false.   Q.E.D.

We leave as an exercise proof of three interesting consequences of the preceding theorem. The third of these, when the restriction to finite sets is removed, has been shown by Tarski [1924a] to be equivalent to the axiom of choice.

THEOREM 47.   *If A is finite and $B \subset A$ then $B < A$.*

THEOREM 48.   *If B and C are finite and if $A \prec B$ and $B \cap C = 0$ then*

$$A \cup C \prec B \cup C.$$

THEOREM 49.   *If A, B, C and D are finite and if $A \prec B$, $C \prec D$ and $B \cap D = 0$ then*

$$A \cup C \prec B \cup D.$$

A property of finite sets closely related to Dedekind finiteness is given in the following theorem.

THEOREM 50.   *If A is finite and $x \notin A$ then $A \prec A \cup \{x\}$.*

We close the systematic development of this section with two useful theorems, the first of which we prove.

THEOREM 51.   *The Cartesian product of two finite sets is finite.*

PROOF.   Let A and B be finite sets. If either A or B is empty, $A \times B = 0$, so we may suppose they are both non-empty. We define:

$$C = \{A \times \{y\} : y \in B\}.$$

We observe that

$$\cup C = A \times B.$$

Moreover, on the basis of Theorem 8

$$A \approx A \times \{y\},$$

whence by virtue of Theorem 42, $A \times \{y\}$ is finite. But $C \approx B$, whence C is finite, and it follows from Theorem 39 that $\cup C$ is finite.   Q.E.D.

THEOREM 52.   *If A and B are finite then A^B is finite.*

Additional properties of finite sets and further equivalent definitions may be obtained by introducing ordering notions, but we shall not pursue these matters here beyond a few informal remarks. For details the reader is again referred to Tarski [1924b]. Stäckel [1907] proposed that finite sets be defined as those which may be doubly well-ordered. More exactly, A is finite if and only if there is a relation R such that both R and $\breve{R}$ well-order A. This definition may be shown to be equivalent to the definition of Tarski used here (see Theorem 37 of Chapter 5).

Some readers may be disappointed that essentially nothing has yet been said about infinite sets. This omission has been deliberate, for until the axiom of infinity is introduced and the set of finite cardinals or finite ordinals shown to exist, little of interest can be conveniently proved about infinite sets. Such sets shall be considered in the latter part of the next chapter.

EXERCISES

1. Give an example different from that in the text of a non-empty family of sets which does not have a minimal or a maximal element.

2. Prove Theorems 24 and 25.

3. Prove Theorems 30 and 31.

4. Give an intuitive example to show that Theorem 33 fails if the hypothesis is omitted that A is finite.

5. Prove Theorem 35.

6. Prove Theorem 36.

7. Prove Theorem 39.

8. Prove Theorems 40 and 41.

9. Prove Theorem 42.

10. Prove Theorem 45.

11. Prove Theorems 47–49.

12. Prove Theorem 50.

13. Prove that any subset of a set which is Dedekind finite is also Dedekind finite.

14. Prove that if $A \approx B$ and A is Dedekind finite then B is Dedekind finite.

15. Prove that if $B \prec A$ and A is Dedekind finite then B is Dedekind finite.

16. Prove Theorem 52.

§ 4.3 Cardinal Numbers.

The Frege-Russell definition of cardinal numbers is beautiful in its simplicity. The cardinal number $\overline{\overline{A}}$ of the set A is the class of all sets equipollent to A, that is,

$$(1) \qquad \overline{\overline{A}} = \{B \colon B \approx A\}.$$

Cantor used the double bar to indicate two levels of abstraction. The first bar is meant to abstract from the particular nature of the elements in the set and the second bar to abstract from their order. (These somewhat vague ideas of abstraction are represented by the clear formal definition of $\overline{\overline{A}}$.) We may, for illustration, define the finite cardinal numbers zero, one and two. (We attach a subscript "f" to indicate we are following Frege's and Russell's original discussion.)

$$0_f = \{0\},$$

that is, 0_f is the set of all sets having no members;

$$1_f = \{A \colon (\exists x)(x \in A \,\&\, (\forall y)(y \in A \to x = y))\},$$

that is, 1_f is the set of all unit sets; an equivalent definition using the relation of equipollence is:

$$1_f = \{A \colon A \approx \{0\}\}.$$

We may similarly proceed to define:

$$2_f = \{A: A \approx \{0, \{0\}\}\},$$

or equivalently:

$$2_f = \{A: (\exists x)(\exists y)(x \in A \; \& \; y \in A \; \& \; x \neq y \; \& \; (\forall z)(z \in A \rightarrow z = x \lor z = y))\}.$$

We then have such results as:

$$\{\text{Edgar Guest, T. S. Eliot}\} \in 2_f.$$

And we see that 2_f is simply the set of all pair sets.

It will also be useful for comparative purposes later to state the classical definition of addition of cardinal numbers. For explicitness, we state:

$$\mathfrak{m} \text{ is a cardinal number} \leftrightarrow (\exists A)(\mathfrak{m} = \overline{\overline{A}}).$$

We then define:

If $\mathfrak{m}$ and $\mathfrak{n}$ are cardinal numbers then

$$(2) \quad \mathfrak{m} + \mathfrak{n} = \{C: (\exists A)(\exists B)(A \in \mathfrak{m} \; \& \; B \in \mathfrak{n} \; \& \; A \cap B = 0 \; \& \; C \approx A \cup B)\}.$$

The intuitive idea of this definition should be clear: the cardinal number which is the *sum* of two cardinal numbers $\mathfrak{m}$ and $\mathfrak{n}$ is the set of all sets which are equipollent to a set consisting of the union of a member of $\mathfrak{m}$ and a member of $\mathfrak{n}$, provided that the two member sets are disjoint. For example, we have:

$$\{\text{C. P. Snow}\} \in 1_f,$$

$$\{\text{Jane Austen, Elizabeth Bowen}\} \in 2_f,$$

and thus

$$2_f + 1_f = 1_f + 2_f = \{A: A \approx \{\text{C. P. Snow}\} \cup$$

$$\{\text{Jane Austen, Elizabeth Bowen}\}\}.$$

It is at once also obvious that

$$2_f + 1_f = 3_f,$$

since

$$\{\text{C. P. Snow, Jane Austen, Elizabeth Bowen}\} \in 3_f.$$

The developments stemming from (1) and (2) are pretty indeed in intuitive set theory, but within our axiomatic framework we cannot show that the set $\overline{\overline{A}}$ is other than the empty set. There are at least three routes we may take to obtain the cardinal numbers in Zermelo-Fraenkel set theory. One is to introduce a new primitive notion and a special axiom for cardinal numbers, which is what we shall do in this chapter. A second is to define cardinal numbers as certain ordinal numbers. This definition requires the axiom of choice to show that every set has a cardinal number; it will be considered in Chapter 8. A third approach is to operate with the present axioms (and later the axiom of infinity) via the notion of the *rank* of a set, but this construction is rather complicated and will not be discussed here.

The special axiom which we introduce requires a new primitive notion, namely, just that of the cardinal number of a set A (in symbols: $\mathcal{K}(A)$). The intuitive idea of the axiom should be transparent. We would like to follow Frege and Russell and define cardinal numbers as equivalence classes of equipollent sets, but we cannot prove that the appropriate equivalence classes exist. So we postulate that with each set A is associated an object $\mathcal{K}(A)$, the cardinal number of A, such that with two equipollent sets we associate the same cardinal number. Notice that on the basis of this axiom and the other axioms introduced we cannot prove that the cardinal number of a set is itself a set. Formally, the *axiom for cardinal numbers* is:

$$\mathcal{K}(A) = \mathcal{K}(B) \leftrightarrow A \approx B.$$

Moreover, in conjunction with the introduction of the new primitive term '$\mathcal{K}$' we need to extend the definition of primitive formula in §2.1 to include this term. This extension also widens the scope of the axiom schema of separation. Any definition or theorem which depends on the new axiom or which uses the new primitive term '$\mathcal{K}$' will be marked with a dagger '†'.*

As far as I know the first explicit use of this new axiom is to be found in Tarski [1924a]; in this article Tarski is concerned to prove that a number of assertions about cardinal numbers are equivalent to the axiom of choice. Naturally for this purpose he needed to construct the cardinal numbers without using the axiom of choice. The systematic developments in the present chapter do not follow Tarski's article, which is concerned with more special and advanced questions than those in which we are interested at this point. One of the best non-axiomatic presentations of the ideas with which we are concerned in this section is to be found in Chapters 2 and 5 of Sierpinski [1928] (see also Sierpinski [1958]).

Our main objective in this section is to develop the elementary arithmetic of cardinal numbers. We shall not distinguish between finite and infinite cardinals at this stage; although a few special theorems about infinite cardinals can be proved, nothing of much interest is possible without the

---

*Note that independent of the axiom we may prove that

(1) $$(\exists x)\,(\mathcal{K}(A) = x),$$

once we admit the primitive term '$\mathcal{K}$', for it is a truth of logic that $\mathcal{K}(A) = \mathcal{K}(A)$, and (1) is a logical consequence of this truth.

axiom of infinity and the set of finite cardinals. In fact, the present section may be viewed as indicating how much of the elementary arithmetic of cardinals is independent of the distinction between finite and infinite sets.

We use German, lower case letters '$\mathfrak{m}$', '$\mathfrak{n}$', '$\mathfrak{p}$', '$\mathfrak{q}$', '$\mathfrak{r}$', with and without subscripts for cardinal numbers. We define for any object x:

†Definition 7. *x is a cardinal number if and only if there is a set A such that $\mathfrak{K}(A) = x$.*

But as in similar situations earlier, we omit the hypothesis 'is a cardinal number' in subsequent theorems and definitions and use the special font of type just indicated. Proofs of most of the theorems follow directly from appropriate theorems in §4.1.

An important tool in the sequel is the theorem asserting that given any two cardinal numbers we can find two mutually exclusive sets corresponding to the two cardinal numbers.

†Theorem 53. *There are sets A and B such that*

   (i) $A \cap B = 0$,

   (ii) $\mathfrak{K}(A) = \mathfrak{m}$,

   (iii) $\mathfrak{K}(B) = \mathfrak{n}$.

proof. In view of Definition 7, there are sets A' and B' such that $\mathfrak{K}(A') = \mathfrak{m}$ and $\mathfrak{K}(B') = \mathfrak{n}$, and by virtue of Theorem 9 there are sets A and B such that $A \cap B = 0$, $A \approx A'$ and $B \approx B'$. By the axiom for cardinals, then, $\mathfrak{K}(A) = \mathfrak{m}$ and $\mathfrak{K}(B) = \mathfrak{n}$. Q.E.D.

We can define addition in a manner very similar to Cantor's, but we first need to prove the appropriate justifying theorem.

†Theorem 54. *There is exactly one cardinal number $\mathfrak{p}$ and there are sets A and B such that*

   (i) $A \cap B = 0$,

   (ii) $\mathfrak{K}(A) = \mathfrak{m}$,

   (iii) $\mathfrak{K}(B) = \mathfrak{n}$,

   (iv) $\mathfrak{K}(A \cup B) = \mathfrak{p}$.

proof. (i)–(iii) follow immediately from the preceding theorem, and the existence of $\mathfrak{p}$ is a truth of logic. We want to show that $\mathfrak{p}$ is independent of the particular sets A and B.

Suppose there were sets A' and B' and a cardinal number $\mathfrak{p}'$ such that

(1) $\qquad\qquad\qquad A' \cap B' = 0$

(2) $\qquad\qquad\qquad \mathfrak{K}(A') = \mathfrak{m}$

(3) $\qquad\qquad\qquad \mathfrak{K}(B') = \mathfrak{n}$

(4) $\qquad\qquad\qquad \mathfrak{K}(A' \cup B') = \mathfrak{p}'$.

It follows from the axiom for cardinals and (2) and (3) that

(5) $$A' \approx A$$

(6) $$B' \approx B.$$

And by virtue of Theorem 4 we infer from (5) and (6) that

$$A' \cup B' \approx A \cup B,$$

whence by the axiom for cardinals

$$\mathcal{K}(A' \cup B') = \mathcal{K}(A \cup B)$$

that is,

$$\mathfrak{p}' = \mathfrak{p},$$

which was to be proved. Q.E.D.

With Theorem 54 at hand, we may define addition of cardinal numbers.

†Definition 8. $\mathfrak{m} + \mathfrak{n} = \mathfrak{p}$ *if and only if there are sets A and B such that*

(i) $A \cap B = 0,$
(ii) $\mathcal{K}(A) = \mathfrak{m},$
(iii) $\mathcal{K}(B) = \mathfrak{n},$
(iv) $\mathcal{K}(A \cup B) = \mathfrak{p}.$

We may easily prove that addition of cardinals is commutative and associative.

†Theorem 55. $\mathfrak{m} + \mathfrak{n} = \mathfrak{n} + \mathfrak{m}.$

Proof. By virtue of Theorem 53 there are sets A and B such that $A \cap B = 0$, $\mathcal{K}(A) = \mathfrak{m}$, and $\mathcal{K}(B) = \mathfrak{n}$, whence by Definition 8

$$\mathcal{K}(A \cup B) = \mathfrak{m} + \mathfrak{n},$$

and

$$\mathcal{K}(B \cup A) = \mathfrak{n} + \mathfrak{m},$$

but

$$A \cup B = B \cup A,$$

whence

$$\mathfrak{m} + \mathfrak{n} = \mathfrak{n} + \mathfrak{m}. \qquad\qquad \text{Q.E.D.}$$

†Theorem 56. $(\mathfrak{m} + \mathfrak{n}) + \mathfrak{p} = \mathfrak{m} + (\mathfrak{n} + \mathfrak{p}).$

Proof. Proof follows from associativity of the union operation on sets. Q.E.D.

We define the cardinal number **0** in the obvious manner, as well as the cardinal numbers 1 and 2.

†Definition 9.

$$0 = \mathcal{K}(0)$$
$$1 = \mathcal{K}(\{0\})$$
$$2 = \mathcal{K}(\{0, \{0\}\}).$$

And we have the theorem, whose simple proof we omit:

†Theorem 57. $\mathfrak{m} + \mathbf{0} = \mathfrak{m}$.

We turn now to multiplication of cardinals. The simple idea is that multiplication of cardinals corresponds to the Cartesian product of sets, which is obviously the case for finite cardinals. Thus if A has 3 elements and B has 4 elements, then $A \times B$ has $3 \cdot 4 = 12$ elements.

Again we need a justifying theorem.

†Theorem 58.   *There is exactly one cardinal number* $\mathfrak{p}$ *and there are sets* A *and* B *such that*

   (i) $\mathcal{K}(A) = \mathfrak{m}$
   (ii) $\mathcal{K}(B) = \mathfrak{n}$
   (iii) $\mathcal{K}(A \times B) = \mathfrak{p}$.

proof. (i)–(ii) and the existence of $\mathfrak{p}$ are immediate. Suppose now there are sets A', B', and a cardinal number $\mathfrak{p}'$ such that

(1) $$\mathcal{K}(A') = \mathfrak{m}$$

(2) $$\mathcal{K}(B') = \mathfrak{n}$$

(3) $$\mathcal{K}(A' \times B') = \mathfrak{p}'.$$

Then (1), (2), and the axiom for cardinals yield:

$$A' \approx A$$
$$B' \approx B,$$

whence by Theorem 5

$$A' \times B' \approx A \times B$$

and thus

$$\mathcal{K}(A' \times B') = \mathcal{K}(A \times B);$$

hence

$$\mathfrak{p}' = \mathfrak{p}. \qquad \text{Q.E.D.}$$

We denote multiplication by juxtaposition as is customary; occasionally a dot is used for purposes of clarity.

†Definition 10.　$mn = p$ *if and only if there are sets A and B such that*

(i)　$\mathcal{K}(A) = m$
(ii)　$\mathcal{K}(B) = n$
(iii)　$\mathcal{K}(A \times B) = p.$

Commutativity and associativity of cardinal multiplication follow at once from the fact that the Cartesian product operation on sets has these properties relative to equipollence.

†Theorem 59.　$mn = nm.$

proof.　Use Theorem 6.

†Theorem 60.　$(mn)p = m(np).$

proof.　Use Theorem 7.

†Theorem 61.　$m \cdot 1 = m.$

proof.　Let

$$\mathcal{K}(A) = m.$$

We know that

$$\mathcal{K}(\{0\}) = 1,$$

and from the definition of multiplication

$$\mathcal{K}(A \times \{0\}) = m \cdot 1,$$

but on the basis of Theorem 8 we know that

$$A \times \{0\} \approx A,$$

whence

$$\mathcal{K}(A \times \{0\}) = m$$

and thus

$$m \cdot 1 = m. \qquad\qquad \text{Q.E.D.}$$

We leave as an exercise proof of the theorem that cardinal multiplication is distributive with respect to cardinal addition.

†Theorem 62.　$m(n + p) = mn + mp.$

The set A^B of all functions from B to A is the basis for defining exponentiation of cardinals. We begin as usual with the justifying theorem, whose proof we omit in this case. The critical theorem to use in the proof is Theorem 10 of §4.1.

†Theorem 63. *There is exactly one cardinal number* $\mathfrak{p}$ *and there are sets* A *and* B *such that*

(i) $\mathfrak{K}(A) = \mathfrak{m}$,
(ii) $\mathfrak{K}(B) = \mathfrak{n}$,
(iii) $\mathfrak{K}(A^B) = \mathfrak{p}$.

†Definition 11. $\mathfrak{m}^\mathfrak{n} = \mathfrak{p}$ *if and only if there are sets* A *and* B *such that*

(i) $\mathfrak{K}(A) = \mathfrak{m}$,
(ii) $\mathfrak{K}(B) = \mathfrak{n}$,
(iii) $\mathfrak{K}(A^B) = \mathfrak{p}$.

By choosing the appropriate theorem from §4.1 we easily prove the following three theorems about exponents.

†Theorem 64. $\mathfrak{m}^{\mathfrak{n}+\mathfrak{p}} = \mathfrak{m}^\mathfrak{n}\mathfrak{m}^\mathfrak{p}$.

†Theorem 65. $(\mathfrak{m}\mathfrak{n})^\mathfrak{p} = \mathfrak{m}^\mathfrak{p}\mathfrak{n}^\mathfrak{p}$.

†Theorem 66. $(\mathfrak{m}^\mathfrak{n})^\mathfrak{p} = \mathfrak{m}^{\mathfrak{n}\mathfrak{p}}$.

Also we may easily prove:

†Theorem 67. $\mathfrak{m}^1 = \mathfrak{m}$.

†Theorem 68. $\mathfrak{m}^0 = 1$.

†Theorem 69.   *If* $\mathfrak{m} \neq 0$ *then* $0^\mathfrak{m} = 0$.

We conclude this section with a few theorems on inequalities among cardinal numbers.

†Definition 12. $\mathfrak{m} \leq \mathfrak{n}$ *if and only if there are sets* A *and* B *such that*

(i) $\mathfrak{K}(A) = \mathfrak{m}$
(ii) $\mathfrak{K}(B) = \mathfrak{n}$
(iii) $A \preceq B$.

And it is easily proved that

†Theorem 70. $\mathfrak{m} \leq \mathfrak{n}$ *if and only if for all* A *and* B *if* $\mathfrak{K}(A) = \mathfrak{m}$ *and* $\mathfrak{K}(B) = \mathfrak{n}$ *then* $A \preceq B$.

From Theorems 1 and 15 it follows at once that:

†Theorem 71. $\mathfrak{m} \leq \mathfrak{m}$.

Using Theorem 17, we may prove:

†Theorem 72.   *If* $\mathfrak{m} \leq \mathfrak{n}$ *and* $\mathfrak{n} \leq \mathfrak{p}$, *then* $\mathfrak{m} \leq \mathfrak{p}$.

And from the Schröder-Bernstein theorem (Theorem 18), we infer that $\leq$ is antisymmetric.

†Theorem 73. *If* $\mathfrak{m} \leq \mathfrak{n}$ *and* $\mathfrak{n} \leq \mathfrak{m}$ *then* $\mathfrak{m} = \mathfrak{n}$.

It is also the case that the three operations introduced are monotonic with respect to the relation $\leq$. The proof follows directly from Theorem 19 of §4.1.

†Theorem 74. *If* $\mathfrak{m} \leq \mathfrak{n}$ *and* $\mathfrak{m}' \leq \mathfrak{n}'$ *then*

(i) $\mathfrak{m} + \mathfrak{m}' \leq \mathfrak{n} + \mathfrak{n}'$,
(ii) $\mathfrak{m}\mathfrak{m}' \leq \mathfrak{n}\mathfrak{n}'$,
(iii) $\mathfrak{m}^{\mathfrak{m}'} \leq \mathfrak{n}^{\mathfrak{n}'}$, *provided it is not the case* $\mathfrak{m} = \mathfrak{m}' = \mathfrak{n} = 0$ & $\mathfrak{n}' \neq 0$.

On the basis of Theorem 20, we have:

†Theorem 75. $\mathfrak{m} \leq \mathfrak{m} + \mathfrak{n}$.

We may define the strict inequality in arithmetical fashion.

†Definition 13. $\mathfrak{m} < \mathfrak{n}$ *if and only if* $\mathfrak{m} \leq \mathfrak{n}$ *and* $\mathfrak{m} \neq \mathfrak{n}$.

We may then prove:

†Theorem 76. $\mathfrak{m} < \mathfrak{n}$ *if and only if there are sets A and B such that*

(i) $\mathcal{K}(A) = \mathfrak{m}$,
(ii) $\mathcal{K}(B) = \mathfrak{n}$,
(iii) $A \prec B$.

We have three obvious properties collected together in the next theorem, which corresponds to Theorem 21.

†Theorem 77.

(i)   *Not* $\mathfrak{m} < \mathfrak{m}$,
(ii) *if* $\mathfrak{m} < \mathfrak{n}$ *then not* $\mathfrak{n} < \mathfrak{m}$,
(iii) *if* $\mathfrak{m} < \mathfrak{n}$ *and* $\mathfrak{n} < \mathfrak{p}$ *then* $\mathfrak{m} < \mathfrak{p}$.

Notice that the assertion that $\mathfrak{m} < \mathfrak{n}$, $\mathfrak{m} = \mathfrak{n}$, or $\mathfrak{n} < \mathfrak{m}$ is simply the law of trichotomy for cardinal numbers. As already pointed out, the equivalence of this law to the axiom of choice will be proved in Chapter 8. What is also important, and surprising, is that if we ask under what circumstances the monotonicity properties expressed in Theorem 74 hold with respect to the strict inequality $<$, the answer, as shown in Tarski [1924a], is that (iii) does not hold at all, and (i) and (ii) are each *equivalent* to the axiom of choice.

An important property which we can prove without the axiom of choice is:

†Theorem 78. $\mathfrak{m} < 2^{\mathfrak{m}}$.

PROOF. Select A so that

$$\mathcal{K}(A) = \mathfrak{m}.$$

Then

$$\mathcal{K}(\{0, \{0\}\}^A) = 2^{\mathfrak{m}}.$$

Moreover, by virtue of Theorem 14

$$\mathcal{P}A \approx \{0, \{0\}\}^A,$$

and in view of Theorem 23

$$A < \mathcal{P}A,$$

whence by Theorem 22

$$A < \{0, \{0\}\}^A,$$

and thus by Theorem 76

$$\mathfrak{m} < 2^{\mathfrak{m}}. \qquad\qquad \text{Q.E.D.}$$

The next theorem asserts that there is no greatest cardinal number. The proof follows the lines of derivation of Cantor's paradox, which was discussed in §1.4.

†THEOREM 79. *For every $\mathfrak{m}$ there is an $\mathfrak{n}$ such that $\mathfrak{m} < \mathfrak{n}$.*

PROOF. Suppose not; that is, suppose there is a greatest cardinal number $\mathfrak{m}$. By virtue of Definition 7 there is then a set A such that

$$\mathcal{K}(A) = \mathfrak{m}.$$

On the other hand, from the axiom for cardinal numbers we know there is an $\mathfrak{n}$ which is the cardinal number of the power set of A, i.e.,

$$\mathcal{K}(\mathcal{P}A) = \mathfrak{n}.$$

In view of Theorem 23

$$A < \mathcal{P}A,$$

and thus by Theorem 76

$$\mathfrak{m} < \mathfrak{n},$$

which is contrary to our supposition.   Q.E.D.

For use in connection with the theory of finite cardinal numbers in the next section, we define the notion of the *successor* $\mathfrak{m}'$ of a cardinal $\mathfrak{m}$. In ordinal number theory the successor of any set A is $A \cup \{A\}$ or $A \cup \{x\}$ for $x \notin A$. A similar approach is suitable here.

†DEFINITION 14. $\mathfrak{m}^{\mathsf{I}} = \mathfrak{n}$ *if and only if there is a set A such that* $\mathcal{K}(A) = \mathfrak{m}$ & $\mathcal{K}(A \cup \{A\}) = \mathfrak{n}$.

We conclude this section with three theorems about the successor function and one more definition.

†THEOREM 80.   *If $\mathfrak{m}^{\mathsf{I}} = \mathfrak{n}^{\mathsf{I}}$ then $\mathfrak{m} = \mathfrak{n}$.*

PROOF.   Let

$$\mathcal{K}(A) = \mathfrak{m}$$
$$\mathcal{K}(B) = \mathfrak{n}.$$

Then by hypothesis

$$A \cup \{A\} \approx B \cup \{B\},$$

under the function f, say.   Then $A \approx B$ under the function g defined by:

$$g = \begin{cases} f|A \text{ if } f(A) = B \\ (f \cup \{\langle f^{-1}(B), f(A)\rangle\}) \sim \{\langle A, f(A)\rangle, \langle f^{-1}(B), B\rangle\} \\ \hspace{5cm} \text{otherwise.} \end{cases}$$

But since $A \approx B$ it follows from the axiom for cardinals that $\mathfrak{m} = \mathfrak{n}$.   Q.E.D.

†THEOREM 81. $\mathfrak{m}^{\mathsf{I}} = \mathfrak{m} + 1$.

†THEOREM 82.   *It is not the case that there is an $\mathfrak{n}$ such that $\mathfrak{m} < \mathfrak{n} < \mathfrak{m}^{\mathsf{I}}$.*

PROOF.   Suppose there were such an $\mathfrak{n}$, and let

$$\mathcal{K}(A) = \mathfrak{m}$$
$$\mathcal{K}(B) = \mathfrak{n}.$$

Then by hypothesis

(1) $A \prec B$

(2) $B \prec A \cup \{A\}.$

Let (2) be established by a function f, say.   There must be a proper subset C of $A \cup \{A\}$ such that

$$B \approx C$$

under f.   Suppose A itself is not in the range of f; then $B \preceq A$, which contradicts (1) and Theorem 22 of §4.1.   Thus A is in the range of f and there must be an element x in $A \sim C$.   Define the function h by:

$$h = (f \cup \{\langle f^{-1}(A), x\rangle\}) \sim \{\langle f^{-1}(A), A\rangle\}.$$

Then clearly

$$B \preceq A$$

under h, and we again contradict (1), so our supposition is absurd.   Q.E.D.

Also for use in the next section we define the set of all predecessors of a cardinal number. Without using the axiom schema of replacement, which is introduced later, we cannot show that for an arbitrary cardinal this set is non-empty, but we can prove this by induction for finite cardinals greater than 0.

†DEFINITION 15. $Q(m) = \{n : n < m\}$

Thus, $Q(0) = 0$ and $Q(1) = \{0\}$.

Every arithmetical theorem stated in this section should be familiar to the reader under the guise of expressing a property of the intuitively given natural numbers. What is important to remember is that these theorems also hold for infinite or transfinite cardinals, and not every familiar property of the natural numbers is shared by infinite cardinals. For example if m is a finite cardinal, then

$$m < m + 1,$$

as would be expected, but if m is a transfinite cardinal

$$m = m + 1.$$

Also the sum of two transfinite cardinals is always equal to or less than their product, which is not true for finite cardinals (since $1 + 2 > 1 \cdot 2$).

In case the reader is puzzled by the use of the phrases 'finite cardinal', 'infinite cardinal', 'transfinite cardinal', it may be said that the topics connoted by these phrases will be explained subsequently. In the next section we discuss the finite cardinals, and in the latter part of the next chapter we distinguish between infinite cardinals and transfinite cardinals; m is an infinite cardinal if there exists an infinite set A such that $\mathcal{K}(A) = m$; m is a transfinite cardinal if there exists a Dedekind infinite set A such that $\mathcal{K}(A) = m$. As is obvious from remarks in the section on finite sets every known proof requires the axiom of choice to show that a cardinal is infinite if and only if it is transfinite.

EXERCISES

1. Prove Theorem 56 in detail.
2. Prove Theorem 57.
3. Prove Theorem 62.
4. Prove Theorem 63.
5. Prove Theorems 64-66.
6. Prove Theorems 67-69.
7. Prove Theorem 70.
8. Prove Theorem 73.
9. Prove Theorem 76.

10. For each of the following, either prove it holds or give a counterexample:
    (a) $\mathcal{K}(A{\sim}B) \leq \mathcal{K}(A)$,
    (b) $\mathcal{K}(A{\sim}B) \leq \mathcal{K}(B)$,
    (c) If $B \subseteq A$ then $\mathcal{K}(A) = \mathcal{K}(A{\sim}B) + \mathcal{K}(B)$,
    (d) $\mathcal{K}(A \cap B) = \mathcal{K}(A)$ if and only if $A \subseteq B$,
    (e) $\mathcal{K}(A \times (B{\sim}C)) = 0$ if and only if $A = 0$ or $B = 0$ or $B = C$.

11. Prove Theorem 81.

12. Consider Definition 15. With the axioms given so far, what difficulty is encountered in trying to prove that $\mathfrak{n} \in \mathcal{Q}(\mathfrak{m})$ if and only if $\mathfrak{n} < \mathfrak{m}$?

§ 4.4 Finite Cardinals.

We now combine the results of the sections on finite sets and cardinal numbers to define the finite cardinal numbers (which we call *finite cardinals* for brevity) and show that they have the expected properties of the intuitively given natural numbers (i.e., non-negative integers). In using the phrase 'intuitively given natural numbers' we do not mean to imply that the natural numbers are well-defined abstract entities which are distinct from, but have many properties in common with, the finite cardinals. Mathematicians have for some time agreed on what the essential properties of natural numbers should be. We show in this section that the finite cardinals — and later the finite ordinals — have these properties, although finite cardinals and ordinals are not necessarily the same entities.

There is, indeed, a crucial distinction between our construction of the cardinals and the ordinals. The ordinals are particular, specified sets, whereas the cardinals are objects which we cannot even classify as sets or individuals. In this respect our position with respect to the character of the cardinals is similar to that of the working mathematician with respect to the natural numbers: he is not concerned with an explicit definition of the natural numbers in terms of other known entities, but only with knowing their essential mathematical properties. In particular, he requires that the natural numbers satisfy Peano's five axioms, which may be formulated in elementary logic independent of set theory. These axioms are based on three primitive symbols: the predicate 'is a natural number', the unary operation symbol '$'$' for the successor function (intuitively $x' = x + 1$), and the individual constant '0' for the number zero. Peano's axioms are then:

P1. *0 is a natural number.*

P2. *If x is a natural number then x' is a natural number.*

P3. *There is no natural number x such that $x' = 0$.*

P4. *If x and y are natural numbers and $x' = y'$ then $x = y$.*

P5. *If $\varphi(0)$ and for every natural number x if $\varphi(x)$ then $\varphi(x')$, then for every natural number x, $\varphi(x)$.*

Axiom P2 says that the successor of any natural number is a natural number. Axiom P3 just says that 0 is the successor of no natural number, that is, it expresses the intuitively obvious fact that there is no natural number x such that

$$x + 1 = 0.$$

Axiom P4 asserts that the successor function is 1–1. Axiom P5 is really an axiom schema which expresses the principle of induction for the natural numbers.

It is a proper question and not an idle philosophical speculation to ask why mathematicians agree almost uniformly on Peano's axioms. Deep and difficult theorems about the natural numbers were proved before any adequate axioms were formulated. Putative axioms which did not yield these theorems would be rejected because, it seems, independent of any axioms there is a quite precise notion of what is true or false of the natural numbers. The author is not prepared to give any exact account of these intuitive notions, and it would be too much of a digression to examine what other people have said about these matters. But it should be realized that to say that the natural numbers *are* the finite cardinals or the finite ordinals is not to give such an exact account, for these intuitive yet precise ideas about the natural numbers are themselves used in deciding if the proposed identification is acceptable.

Proceeding now to formal developments, we define finite cardinals as cardinals of finite sets — where by finite set we mean a finite set in the sense of Tarski.*

†DEFINITION 16. *x is a finite cardinal if and only if there is a finite set A such that* $\mathcal{K}(A) = x$.†

We begin by proving Peano's axioms for the finite cardinals.

†THEOREM 83. **0** *is a finite cardinal.*

PROOF. Immediate from Definition 9 and Theorem 24.

†THEOREM 84. *If* m *is a finite cardinal then* m' *is a finite cardinal.*

PROOF. On the basis of the hypothesis that m is a finite cardinal, there is a finite set A such that $\mathcal{K}(A) = $ m, and by virtue of the definition of the successor function

$$\mathcal{K}(A \cup \{A\}) = \text{m}'$$

---

*I have not seen the theory of finite cardinals developed on this basis in the literature, but it seems to be a highly natural approach.

†We do not introduce a special font of type for finite cardinals, for we reserve the lower-case italic letters '*m*', '*n*', '*p*', '*q*', '*r*' for finite ordinals. The reason for this is that in Chapter 6 we construct the rational and real numbers from the finite ordinals rather than the finite cardinals in order not to make this construction depend on the axiom for cardinal numbers.

and it follows from the fact that A is finite and Theorem 29 that $A \cup \{A\}$ is finite, whence $\mathfrak{m}^{\mathfrak{l}}$ is a finite cardinal. Q.E.D.

†THEOREM 85. *There is no finite cardinal $\mathfrak{m}$ such that $\mathfrak{m}^{\mathfrak{l}} = 0$.*

PROOF. Suppose by way of contradiction there were such an $\mathfrak{m}$. Let $\mathcal{K}(A) = \mathfrak{m}$. Then from the definition of 0 and successor, and the axiom for cardinals

$$A \cup \{A\} \approx 0,$$

which is absurd. Q.E.D.

The proof of Peano's axiom P4 for finite cardinals is immediate from Theorem 80 and need not be stated.

Finally, to prove the principle of induction for finite cardinals we use the principle of induction for finite sets (Theorem 33).

†THEOREM SCHEMA 86. *If*

(i) $\varphi(0)$,
(ii) *for every finite cardinal $\mathfrak{m}$ if $\varphi(\mathfrak{m})$ then $\varphi(\mathfrak{m}^{\mathfrak{l}})$,*
then for every finite cardinal $\mathfrak{m}$, $\varphi(\mathfrak{m})$.

PROOF. Suppose that there were a finite cardinal $\mathfrak{m}$ for which $\varphi(\mathfrak{m})$ is false. Let A be a finite set such that $\mathcal{K}(A) = \mathfrak{m}$. We derive a contradiction by induction on subsets of A. Define:

$$L = \{B : B \subseteq A \ \& \ \varphi(\mathcal{K}(B))\}.$$

From (i) of the hypothesis we know that $0 \in L$. Assume now for the induction that $B \in L$ and $x \in A \sim B$. From the definition of L the set B is finite because it is a subset of the finite set A. We want to show that $\varphi(\mathcal{K}(B \cup \{x\}))$. We observe first that obviously

(1) $$B \cup \{B\} \approx B \cup \{x\}.$$

Let $\mathcal{K}(B) = \mathfrak{n}$. Then by (ii) of the hypothesis and the fact that $B \in L$ and thus $\varphi(\mathcal{K}(B))$, we infer $\varphi(\mathfrak{n}^{\mathfrak{l}})$, but

(2) $$\mathcal{K}(B \cup \{B\}) = \mathfrak{n}^{\mathfrak{l}},$$

whence from (1) and (2) and the axiom for cardinals

$$\mathcal{K}(B \cup \{x\}) = \mathfrak{n}^{\mathfrak{l}}$$

and thus $\varphi(\mathcal{K}(B \cup \{x\}))$ and $B \cup \{x\} \in L$. It then follows from Theorem 33 that $A \in L$ and thus $\varphi(\mathcal{K}(A))$, that is, $\varphi(\mathfrak{m})$, which contradicts our supposition. Q.E.D.

By building on these five theorems and the general development of cardinal arithmetic in §4.3, the complete elementary arithmetic of addition, multiplication, and exponentiation is easily constructed.

We state without proof four theorems about the relation *less than*. Proofs are immediate from facts about finite sets proved in §4.2.

†Theorem 87.   *If* $\mathfrak{m}$ *is a finite cardinal then* $\mathfrak{m} < \mathfrak{m} + 1$.

†Theorem 88.   *If* $\mathfrak{m}$ *is a finite cardinal and* $\mathfrak{n}$ *is not, then* $\mathfrak{m} < \mathfrak{n}$.

†Theorem 89.   *If* $\mathfrak{m}$ *is a finite cardinal and* $\mathfrak{n} < \mathfrak{m}$ *then* $\mathfrak{n}$ *is a finite cardinal.*

†Theorem 90.   *If* $\mathfrak{m}$ *is a finite cardinal then* $\mathfrak{n} < \mathfrak{m}, \mathfrak{n} = \mathfrak{m}$, *or* $\mathfrak{m} < \mathfrak{n}$.

We conclude this section with a sequence of theorems leading to the result that $<$ well-orders any set of finite cardinals. We first prove by induction that the set $\mathcal{Q}(\mathfrak{m})$ of predecessors of a finite cardinal $\mathfrak{m}$ is non-empty if $\mathfrak{m}$ is greater than $\mathbf{0}$.

†Theorem 91.   *If* $\mathfrak{m}$ *is a finite cardinal then* $\mathfrak{n} \in \mathcal{Q}(\mathfrak{m})$ *if and only if* $\mathfrak{n} < \mathfrak{m}$.

proof.   From the character of definition by abstraction, which was used in defining $\mathcal{Q}(\mathfrak{m})$, it is clear that to establish the theorem we need to prove that for every $\mathfrak{m}$ there is a set A such that for every $\mathfrak{n}$, $\mathfrak{n} \in A$ if and only if $\mathfrak{n} < \mathfrak{m}$. For $\mathfrak{m} = 0$, we may take $A = 0$. For the second part of the induction, we assume there is such a set A for $\mathfrak{m}$. But it is easy to see that

$$B = A \cup \{\mathfrak{m}\}$$

is the appropriate set for $\mathfrak{m} + 1$, that is, on the inductive hypothesis for A, $\mathfrak{n} \in B$ if and only if $\mathfrak{n} < \mathfrak{m} + 1$. Q.E.D.

It is also not difficult to prove that if $\mathfrak{m}$ is a finite cardinal then $\mathcal{Q}(\mathfrak{m})$ has $\mathfrak{m}$ members.

†Theorem 92.   *If* $\mathfrak{m}$ *is a finite cardinal then* $\mathcal{K}(\mathcal{Q}(\mathfrak{m})) = \mathfrak{m}$.

We next want to prove that $<$ well-orders $\mathcal{Q}(\mathfrak{m})$, when $\mathfrak{m}$ is a finite cardinal. As matters now stand, the symbol '$<$' does not designate a set, just as '$=$' and '$\subseteq$' do not, but by appropriate restriction, along the lines we used for identity in Chapter 2, the desired set-theoretical entity may be obtained.

†Definition 17. $< | A = \{\langle \mathfrak{n}, \mathfrak{m} \rangle : \mathfrak{n} \in A \ \& \ \mathfrak{m} \in A \ \& \ \mathfrak{n} < \mathfrak{m}\}$.
(Notice that we have used the vertical bar of restriction defined in Chapter 3 because of its suggestiveness, although actually here the whole symbol '$<|$' is a unary operation symbol.) It is a simple matter to apply the axiom schema of separation to obtain:

†Theorem 93. $\mathfrak{n} < | A \ \mathfrak{m}$ *if and only if* $\mathfrak{n} \in A \ \& \ \mathfrak{m} \in A \ \& \ \mathfrak{n} < \mathfrak{m}$.

To avoid cumbersome notation, it is convenient to define:

†Definition 18.　$<_{\mathfrak{m}} = <|\,\mathcal{Q}(\mathfrak{m})$.

We may then prove by induction (using Theorem 86) that

†Theorem 94.　*If* $\mathfrak{m}$ *is a finite cardinal then* $<_{\mathfrak{m}}$ *well-orders* $\mathcal{Q}(\mathfrak{m})$.

proof.　Obviously $<_0$ well-orders $\mathcal{Q}(0)$, since $\mathcal{Q}(0)$ is empty.

For the second part of the induction, we assume that

(1) $$<_{\mathfrak{m}} \text{ well-orders } \mathcal{Q}(\mathfrak{m}).$$

Now it is obvious from the relevant definitions that

$$\mathcal{Q}(\mathfrak{m}') = \mathcal{Q}(\mathfrak{m}) \cup \{\mathfrak{m}\}.$$

We need to show that any non-empty subset A of $\mathcal{Q}(\mathfrak{m}')$ has an $<_{\mathfrak{m}'}$ - first element, for the connectivity of $\mathcal{Q}(\mathfrak{m}')$ under the given relation follows from Theorem 90. (These are the two conditions required for $<_{\mathfrak{m}'}$ to well-order $\mathcal{Q}(\mathfrak{m}')$; cf. Definition 28 of Chapter 3.) If $A \subseteq \mathcal{Q}(\mathfrak{m})$ then that A has a $<_{\mathfrak{m}'}$ - first element follows from the inductive hypothesis (1). At the other extreme, if $A = \{\mathfrak{m}\}$, the desired conclusion follows at once. The third possibility, i.e., that not $A \subseteq \mathcal{Q}(\mathfrak{m})$ and $A \cap \mathcal{Q}(\mathfrak{m}) \neq 0$, is equally simple. For $A \sim \{\mathfrak{m}\} \subseteq \mathcal{Q}(\mathfrak{m})$ and thus $A \sim \{\mathfrak{m}\}$ has a $<_{\mathfrak{m}'}$ - first element, say $\mathfrak{n}$, and clearly $\mathfrak{n} < \mathfrak{m}$, whence $\mathfrak{n}$ is also the $<_{\mathfrak{m}'}$ - first element of A, which completes our inductive proof. Q.E.D.

Finally, we prove:

†Theorem 95.　*Any set* A *of finite cardinals is well-ordered by* $<|A$.

proof.　If A is empty there is nothing to prove, so we may assume that $A \neq 0$. Connectivity of $<|A$ follows from Theorem 90. Let B be a non-empty subset of A. To complete our proof we need to show that B has a $<|A$ - first element. Choose an element $\mathfrak{m}$ from B. If $B \cap \mathcal{Q}(\mathfrak{m}) = 0$ then clearly $\mathfrak{m}$ is the $<|A$ - first element of B. If $B \cap \mathcal{Q}(\mathfrak{m}) \neq 0$, let

$$C = B \cap \mathcal{Q}(\mathfrak{m}).$$

Obviously if $\mathfrak{n} \in C$ and $\mathfrak{p} \in B \sim C$

(1) $$\mathfrak{n} < \mathfrak{p},$$

but since C is a non-empty subset of $\mathcal{Q}(\mathfrak{m})$, by virtue of the preceding theorem C has a $<_{\mathfrak{m}}$ - first element, say $\mathfrak{n}^*$, and in view of (1) there is no difficulty in showing that $\mathfrak{n}^*$ is the $<|A$-first element of B. Q.E.D.

It may be remarked that there is no hope of proving without the axiom of choice the analogue of Theorem 95 for a set of arbitrary cardinals, for the connectivity of the set is simply the law of trichotomy.

EXERCISES

1. Prove Theorems 87 and 88.
2. Prove Theorems 89 and 90.
3. Prove Theorems 92 and 93.

In the next eight exercises, assume that m, n, p *and* q *are finite cardinals.*

4. Prove that if $m + n = p$ then $m \leq p$.
5. Prove that $n \leq m + n$.
6. Prove that if $m < n$ then $m + p < n + p$.
7. Prove that if $m + p < n + p$ then $m < n$.
8. Prove that if $m < n$ & $p < q$ then $m + p < n + q$.
9. Prove that if $m < n$ & $p \neq 0$ then $mp < np$.
10. Prove that if $mp < np$ then $m < n$.
11. Prove that if $m < n$ & $p < q$ then $mp < nq$.
12. Let R be a relation whose field is a set of finite cardinals. Prove that there is a function f with $\mathfrak{D}R = \mathfrak{D}f$ and $f \subseteq R$. (This assertion, with the restriction to finite cardinals removed, is one form of the axiom of choice.)

CHAPTER 5

FINITE ORDINALS AND DENUMERABLE SETS

§ 5.1 Definition and General Properties of Ordinals. In this chapter we first develop to a certain extent the general theory of ordinal numbers, following the ideas of von Neumann [1923]. We then turn our attention to the finite ordinals whose theory we develop independently of that of finite cardinals in the final section of the previous chapter and also independently of the special axiom for cardinal numbers. At no point in this or subsequent chapters does the theory of ordinal numbers depend on this axiom. In the last section of this chapter we consider the theory of denumerable sets and return briefly to the theory of cardinal numbers.

The theory of finite ordinals given in this chapter is certainly more complicated than that of the finite cardinals in Chapter 4, but it has the great virtue of not depending on any special axiom. When the theory of cardinal numbers is made independent of this special axiom and the axiom of choice by use of the notion of the rank of a set (cf. Scott [1955]), it is, in its opening phases, more complicated than the theory of ordinals. This approach, via the concept of rank, will not be pursued here.

We begin with some description of the classical Cantor theory of ordinal numbers. This theory depends upon first defining the notion of *order type* and then designating ordinal numbers as certain special order types, namely, the order types of well-ordered sets. Loosely speaking, an order type is the set of all simply ordered sets which can be put into 1–1 correspondence with a given simply ordered set such that the order is "preserved." In making this more precise, it is desirable in the interest of clarity to deal with ordered pairs: $\mathfrak{A} = \langle A, R \rangle$ is a *simple order structure* if and only if R is a strict simple ordering of A. Classically, reference to the ordering R is suppressed and a single bar over A is used to indicate the order type. We shall be more explicit. First, we define the notion of two simple order structures being *similar* to catch the notion of an order preserving 1–1 correspondence. The definition does not need the assumption that the pairs are simple order structures; a format common in abstract algebra is followed, which explains the use of German letter variables '$\mathfrak{A}$' and '$\mathfrak{B}$'. (This definition is a special case of the general definition of isomorphism of algebraic systems.)

DEFINITION 1.

(i) $\mathfrak{A} = \langle A, R \rangle$ is similar under f to $\mathfrak{B} = \langle B, S \rangle$ if and only if

    (a) f is a 1-1 function,

    (b) $\mathfrak{D}f = A$ and $\mathfrak{R}f = B$,

    (c) for every $x, y \in A$

$$xRy \text{ if and only if } f(x) S f(y).$$

(ii) $\mathfrak{A} = \langle A, R \rangle$ is similar to $\mathfrak{B} = \langle B, S \rangle$ if and only if there is an f such that $\mathfrak{A}$ is similar under f to $\mathfrak{B}$.

Obviously the notation: $\mathfrak{A} = \langle A, R \rangle$ in the definiendum does not satisfy our formal rules of definition. We should have:

$\mathfrak{A}$ is similar under f to $\mathfrak{B}$ if and only if there are sets A, B, R, and S such that $\mathfrak{A} = \langle A, R \rangle$, $\mathfrak{B} = \langle B, S \rangle$ and . . .

But the usage we have followed is widespread and intuitively more perspicuous. If, for instance,

$$A = \{1, 2, 3\}$$
$$B = \{3, 5, 7\}$$
$$R = \text{ less than restricted to } A$$
$$S = \text{ greater than restricted to } B,$$

then $\langle A, R \rangle$ is similar to $\langle B, S \rangle$, for we may take as the appropriate function:

$$f = \{\langle 1, 7 \rangle, \langle 2, 5 \rangle, \langle 3, 3 \rangle\}.$$

In intuitive set theory we would, if it were not for the paradoxes, define: if $\mathfrak{A}$ is a simple order structure then

(1) $$\overline{\mathfrak{A}} = \{\mathfrak{B} : \mathfrak{B} \text{ is similar to } \mathfrak{A}\}.$$

The difficulty with (1) is exactly the difficulty we encountered with the corresponding definition of cardinal numbers: we cannot prove that the equivalence class $\overline{\mathfrak{A}}$, which is the order type of $\mathfrak{A}$, is not empty. Easy and natural development of intuitive order type theory would require the introduction of a new axiom like the axiom for cardinals.*

---

*We may mention the four most important order types and their commonly agreed upon symbolic names. Thus

    ω is the order type of $\langle N, < \rangle$, where N is the set of natural numbers;

    ω^* is the order type of $\langle N, > \rangle$, or just as well, the order type of the negative integers under *less than*;

    η is the order type of $\langle \text{Rat}, < \rangle$, where Rat is the set of rational numbers;

    λ is the order type of $\langle \text{Re}, < \rangle$, where Re is the set of real numbers. Each of these order types may be characterized abstractly (see, for instance, Sierpinski [1928]).

It should be mentioned that order types may be defined simply in terms of relations. If R is a simple ordering, then

$$\overline{R} = \{S : \langle \mathfrak{F}S, S \rangle \text{ is similar to } \langle \mathfrak{F}R, R \rangle\},$$

but none of the difficulties of (1) are avoided by this approach.

In view of the difficulties with definition by abstraction of cardinal numbers or order types, it is fortunate indeed that for the special case of ordinal numbers, which are classically order types of well-ordered sets, a device may be adopted which requires no special axioms to support it. The idea is to choose just one representative of each order type which is an ordinal and call this representative *the* ordinal. That is, in the case of well-orderings we are actually able to construct a definite example of each possible well-ordering, which we cannot do in the case of arbitrary orderings.

The definite representatives we choose are built up from the empty set:

$$\mathbf{1} = \{0\}$$
$$\mathbf{2} = \{0, \{0\}\} = \{0, 1\}$$
$$\mathbf{3} = \{0, \{0\}, \{0, \{0\}\}\} = \{0, 1, 2\}$$
$$\cdot \ \cdot \ \cdot \ \cdot \ \cdot \ \cdot \ \cdot \ \cdot \ \cdot \ \cdot \ \cdot \ \cdot \ \cdot \ \cdot \ \cdot$$

Each ordinal has all smaller ordinals as members, and the membership relation provides the appropriate well-ordering. This construction is due to von Neumann [1923]. Its intuitive adequacy to provide a representative of each type of well-ordering may be explained by the following informal argument, for which I am indebted to Dana Scott.

Let R be a well-ordering of A. We want to give a method of associating a new well-ordering relation with R that does not depend on the particular nature of the elements of A. Put another way, we want to define a function, say f, on A which will yield an appropriate definite representative matching the ordering due to R. The definition proceeds as follows. Let a_0 be the R-first element of A. We set

$$f(a_o) = 0.$$

Let a_1 be the next element of A under R. We set

$$f(a_1) = \{f(a_o)\}$$
$$= \{0\}.$$

Similarly, for a_2

$$f(a_2) = \{f(a_o), f(a_1)\}$$
$$= \{0, \{0\}\},$$

and for any integer n

$$f(a_n) = \{f(a_o), f(a_1), \ldots, f(a_{n-1})\}.$$

Moreover, this construction may be extended beyond the integers. In general, the value of f for each element a is simply the set of function values of the elements which precede a in the ordering R. Formally, for every a in A

$$f(a) = f``\breve{R}``\{a\}.$$

Moreover, it is not difficult to show that for $a, b \in A$

$$aRb \text{ if and only if } f(a) \in f(b).$$

Further, if S well-orders B and $\langle A, R \rangle$ is similar to $\langle B, S \rangle$ under a function g say, then for a in A

$$f(a) = f'(g(a)),$$

where f' is constructed for $\langle B, S \rangle$ as f was for $\langle A, R \rangle$. The functions f and f' have mapped the two well-orderings onto an appropriate representative which replaces the relations R and S by a fragment of the membership relation. Thus instead of definition by abstraction we have a simple and natural procedure for representing any well-ordering by the membership relation. In Chapter 7 (Theorem 81) we prove that this representation is indeed always possible.

We now proceed to formal construction of the ordinals. Some preliminary notions are needed. We begin with the definition of "fragments" of the membership relation. We define a relation $\mathcal{E}A$ which corresponds to membership exactly as $\mathcal{I}A$ corresponds to identity.

DEFINITION 2. $\mathcal{E}A = \{\langle x, y \rangle \colon x \in A \ \& \ y \in A \ \& \ x \in y\}$.

Mainly by virtue of the axiom schema of separation we have:

THEOREM 1. $\langle x, y \rangle \in \mathcal{E}A$ *if and only if* $x \in A \ \& \ y \in A \ \& \ x \in y$.

DEFINITION 3. A *is complete if and only if every member of* A *is a subset of* A.

Some examples will illustrate this notion. Let

$$A_1 = \{\text{James Joyce}, \{0\}\}.$$

$$A_2 = \{0, \{0\}\}.$$

Then A_2 is complete, but A_1 is not. Notice that we could not define completeness by the condition:

(1)   For every x if $x \in A$ then $x \subseteq A$,

because the definition of '$\subseteq$' was conditional and no decision can be made about the relation of inclusion for individuals. For example, with (1) as the definiens we would not be able to decide if {James Joyce} is complete.

The English wording of Definition 3 is intended to entail that every member of a complete set is a set.

We have the simple result:

THEOREM 2. *If A and B are complete then $A \cap B$ and $A \cup B$ are complete.*

We now can state the definition of ordinals which in its present simple form is due to Robinson [1937].

DEFINITION 4. *A is an ordinal if and only if A is complete and $\mathcal{E}A$ connects A.*

The adequacy of this definition depends on the axiom of regularity. Without the assumption of that axiom we need something like the following: A is an ordinal if and only if

(i) $\mathcal{E}A$ well-orders A,
(ii) A is complete.

Condition (i) prohibits an infinitely descending sequence of elements in A, which is also prohibited by the axiom of regularity. And without this axiom, we have $0 \neq \{0\}$, $\{0\} \neq \{0, \{0\}\}$, etc.

In the remainder of this section we prove some general theorems about ordinals, most of which are necessary to the construction of the finite ordinals in the following section. We begin by proving that every ordinal is well-ordered by the membership relation. As would be expected from the above remarks, the proof depends on the axiom of regularity.

THEOREM 3. *If A is an ordinal then $\mathcal{E}A$ well-orders A.*

PROOF. Since if A is an ordinal $\mathcal{E}A$ connects A, we need show only that every non-empty subset B of A has an $\mathcal{E}A$-first element. By the axiom of regularity and the fact that every member of B is a set, there is a set C in B such that

$$B \cap C = 0.$$

Hence no member of B is also a member of C, and thus C is an $\mathcal{E}A$-first member of B.   Q.E.D.

The proof of the next theorem uses facts about $\mathcal{E}A$-sections established in §3.2.

THEOREM 4. *If A is an ordinal, $B \subset A$, and B is complete, then $B \in A$.*

PROOF. In view of Theorem 3 we know that $\mathcal{E}A$ well-orders A. Moreover, since B is complete, if $x \mathcal{E}A y$ and $y \in B$, then $x \in B$. Hence B is an $\mathcal{E}A$-section of A and thus by virtue of Theorem 69 of §3.2 there is a z in A such that

(1) $B = \{x: x \in A \;\&\; x \mathcal{E}A z\}.$

But since A is complete, every member of z is a member of A and (1) reduces simply to:

$$B = \{x\colon x \in z\},$$

that is,

$$B = z,$$

and z is a member of A. Q.E.D.

THEOREM 5. *If A and B are ordinals, then $A \subset B$ if and only if $A \in B$.*

PROOF. If $A \subset B$ then since A is complete, by the preceding theorem, $A \in B$. If $A \in B$ then since B is complete $A \subset B$. Q.E.D.

THEOREM 6. *Every member of an ordinal is an ordinal.*

PROOF. Let A be an ordinal and $B \in A$. Since A is complete, $B \subseteq A$ and thus $\mathcal{E}B$, which is a subset of $\mathcal{E}A$, connects B. Moreover, since $\mathcal{E}A$ well-orders A, it well-orders B, and hence is transitive on B. Thus if we have:

$$x \ \mathcal{E}A \ y \ \& \ y \ \mathcal{E}A \ B,$$

we infer $x \ \mathcal{E}A \ B$, that is $x \in B$, whence $y \subseteq B$, and B is complete. Hence B is an ordinal. Q.E.D.

Proofs of the next two theorems are left as exercises. The first one is slightly difficult.

THEOREM 7. *If A and B are ordinals then either $A \subseteq B$ or $B \subseteq A$.*

THEOREM 8. *If A and B are ordinals then exactly one of the following holds*:

$$A \in B, B \in A, A = B.$$

THEOREM 9. *If B is a set of ordinals then $\cup B$ is an ordinal.*

PROOF. Since B is a set of ordinals and members of ordinals are ordinals (preceding theorem), in view of Theorem 8, $\mathcal{E} \cup B$ connects $\cup B$. To show that $\cup B$ is complete, let $C \in \cup B$. Then there is an ordinal D such that $C \in D$ and $D \in B$. Since D is complete, $C \subseteq D$, and by virtue of Theorem 62 of §2.6 from $D \in B$ we infer $D \subseteq \cup B$, and thus by transitivity of inclusion we conclude: $C \subseteq \cup B$. Q.E.D.

This proof illustrates the typical, straightforward technique for proving that some set is an ordinal: prove that the set is connected by the membership relation and that it is complete.

We define strict and weak *less than* for sets in terms of membership, and also strict and weak *greater than*.

DEFINITION 5.

(i) $A < B$ *if and only if* $A \in B$,
(ii) $A \leq B$ *if and only if* $A < B$ *or* $A = B$,
(iii) $A > B$ *if and only if* $B < A$,
(iv) $A \geq B$ *if and only if* $B \leq A$.

We restrict ourselves to one theorem about strict less than, but the elementary order properties of the other three relations are easily proved, and we assume them in the sequel.

THEOREM 10.   *If A, B, and C are ordinals then*

(i)   *not* $A < A$,
(ii) *if* $A < B$ *then not* $B < A$,
(iii) *if* $A < B$ *and* $B < C$ *then* $A < C$,
(iv) *exactly one of the following holds*:

$$A < B, A = B, B < A.$$

The next theorem shows that each ordinal is just the set of smaller ordinals. We omit the proof.

THEOREM 11.   *If A is an ordinal then*

$$A = \{B : B \text{ is an ordinal } \& B < A\}.$$

We now prove that the set of all ordinals does not exist.

THEOREM 12.   *There is no set A such that for every x, $x \in A$ if and only if x is an ordinal.*

PROOF. The proof consists in showing that if there were such a set, say A, then it would be an ordinal and thus a member of itself, violating the fact that $\mathcal{E}A$ well-orders A. That is, it is absurd to have $A \in A$.

Suppose that A is the set of all ordinals. By virtue of Theorem 8, $\mathcal{E}A$ connects A. Let B be an arbitrary member of A. In view of Theorem 6 every member of B is an ordinal, whence a member of A, and thus $B \subseteq A$, which establishes the completeness of A. A is therefore an ordinal, but then $A \in A$.   Q.E.D.

This theorem shows that in Zermelo set theory the Burali-Forti paradox of the greatest ordinal (cf. the discussion in §1.4) is blocked by the fact that the set of all ordinals does not exist. It is to be emphasized that the proof of this theorem does not really depend on the axiom of regularity, for if this axiom is not assumed, then as already indicated, it is part of the definition of ordinals to require that if A is an ordinal $\mathcal{E}A$ well-orders A, which implies $A \notin A$.*

*In von Neumann set theory the class of all ordinals does exist, but it is a proper class and thus cannot be a member of itself or any other class.

We conclude this section by introducing the notion of *successor* for ordinals. The intuitive idea is the same as for cardinals and the definitions are closely related. No confusion will result from using the same standard symbol for both successor functions, for in every context of occurrence it will be clear whether cardinals (constructed by use of the special axiom for cardinals) or ordinals are being considered.

DEFINITION 6. $A^{\mathsf{I}} = A \cup \{A\}$.

For example,

$$0^{\mathsf{I}} = 0 \cup \{0\} = \{0\} = 1$$
$$1^{\mathsf{I}} = \{0\}^{\mathsf{I}} = \{0\} \cup \{\{0\}\} = \{0, \{0\}\} = 2.$$

We state without proof some theorems very similar to theorems already stated for cardinals.

THEOREM 13. *If* $A^{\mathsf{I}} = B^{\mathsf{I}}$ *then* $A = B$.

THEOREM 14. *If* A *is an ordinal then* $\cup A^{\mathsf{I}} = A$.

THEOREM 15. *If* A *is an ordinal then there is no ordinal* B *such that*

$$A < B < A^{\mathsf{I}}.$$

THEOREM 16. A^{I} *is an ordinal if and only if* A *is an ordinal.*

EXERCISES

1. Prove that Definition 4 of ordinals is equivalent to the other definition given immediately after it in the text.

2. Give a detailed proof of Theorem 1.

3. Prove Theorem 2.

4. Prove that a set A whose only members are sets is complete if and only if $\cup A \subseteq A$.

5. Prove Theorems 7 and 8. (Hint: Suppose $A \cap B \subset A$ & $A \cap B \subset B$ and then derive an absurdity.)

6. Prove Theorem 10.

7. Prove Theorem 11.

8. Prove Theorems 13 and 14.

9. Prove Theorems 15 and 16.

10. Prove that if A and B are ordinals and $A \in B$ then $B \notin A^{\mathsf{I}}$.

11. Let B be a set of ordinals. Prove
    (a) If $C \in B$ then $C \leq \cup B$,
    (b) If D is an ordinal and for every C in B, $C \leq D$, then $\cup B \leq D$.

12. Give a counterexample to show that (b) Exercise 11 is false if the requirement is omitted that D be an ordinal.

13. Prove that if B is a non-empty set of ordinals then $\cap B$ is the $\mathcal{E}B$-first member of B.

§ 5.2 Finite Ordinals and Recursive Definitions.

Finite ordinals are simply ordinals which are well-ordered by the converse of the membership relation. For intuitive ease of reference in constructing the real numbers in the next chapter we call the finite ordinals *natural numbers*.

> **DEFINITION 7.** *A is a natural number if and only if A is an ordinal and* $\breve{\varepsilon}A$ *well-orders A.*

This definition could be replaced by the following: *A* is a natural number if and only if *A* is an ordinal and *A* is a finite set in the sense of Tarski. As a matter of fact, to indicate how quickly and directly the theory of natural numbers as finite ordinals may be developed from the axioms of Zermelo set theory, we have made most of the present section independent of the theory of finite sets given in §4.2.

For natural numbers we shall use the lower-case italic letters '*m*', '*n*', '*p*', '*q*', '*r*', '*s*', and '*t*', with and without subscripts.

Our first task is to prove Peano's five axioms (see §4.4). The proof is obvious that

> **THEOREM 17.** *0 is a natural number.*

> **THEOREM 18.** *If n is a natural number then n^{I} is a natural number.*

PROOF. Since *n* is a natural number, *n* is an ordinal, and thus by the last theorem of the previous section n^{I} is an ordinal. Whence we know that εn^{I} and thus $\breve{\varepsilon}n^{\text{I}}$ connect n^{I}. To show that n^{I} is well-ordered by $\breve{\varepsilon}n^{\text{I}}$ and is consequently a natural number, we need prove only that every non-empty subset of n^{I} has an $\breve{\varepsilon}n^{\text{I}}$-first element. If a non-empty subset of n^{I} has *n* as a member, *n* is its $\breve{\varepsilon}n^{\text{I}}$-first member, and if it does not, it is a subset of *n* and by hypothesis of the theorem has an $\breve{\varepsilon}n$-first element and thus an $\breve{\varepsilon}n^{\text{I}}$-first element.   Q.E.D.

> **THEOREM 19.** *There is no natural number n such that $n^{\text{I}} = 0$.*

PROOF. Suppose there were such a natural number *n*. Then $n \in n^{\text{I}}$, whence $n \in 0$, which is absurd.   Q.E.D.

> **THEOREM 20.** *If n and m are natural numbers and $n^{\text{I}} = m^{\text{I}}$ then $n = m$.*

PROOF. Immediate consequence of Theorem 13.   Q.E.D.

Proof of the induction schema is facilitated by the following theorem, whose proof we leave as an exercise.

> **THEOREM 21.** *If $A \leq n$ then A is a natural number.*

We now prove:

THEOREM SCHEMA 22. *If*

   (i) $\varphi(0)$,

   (ii) *for every n if $\varphi(n)$ then $\varphi(n^{\text{I}})$*,

then for every n, $\varphi(n)$.

PROOF. Assume the hypothesis of the theorem and suppose there is an n such that it is not the case $\varphi(n)$. Let

$$L(n) = \{B: B \leq n \;\&\; \varphi(B) \text{ is false}\}.$$

Then $\mathcal{E}n^{\text{I}}$ well-orders $L(n)$. Let B^* be the first element of $L(n)$. Since B^* is not zero, $\widetilde{\mathcal{E}B^*}$ well-orders B^* and there is an $\widetilde{\mathcal{E}B^*}$-first element of B^*, say D. That is, D is the $\mathcal{E}B^*$-last element of B^*. Moreover, because $D < B^*$

  (1) $\varphi(D)$.

But

  (2) $D^{\text{I}} = B^*$,

since if $D^{\text{I}} \neq B^*$, then

$$D^{\text{I}} \in B^* \qquad\qquad \text{(Why?)}$$

and D would not be the $\mathcal{E}B^*$-last element of B^*. Whence from (1), (2) and the hypothesis of the theorem we infer that

$$\varphi(B^*),$$

which is absurd. Q.E.D.

Theorems 17-20 and Theorem 22 correspond to Peano's five axioms.

The further course of development of the arithmetic of the natural numbers cannot proceed along the lines that were followed for the arithmetic of the finite cardinals. The axiom for cardinals made direct justification of the explicit definitions of cardinal addition, multiplication, and exponentiation simple and natural.

To illustrate the problems which face us, we may for the moment concentrate on defining addition. Given simply Peano's axioms P1–P5 formulated in predicate logic with identity and without set theory, it may be shown (J. Robinson [1949]) that a proper, explicit definition of addition cannot be stated within this framework. The customary procedure is to adopt two further axioms:

P6.   *If x is a natural number then $x + 0 = x$.*

P7.   *If x and y are natural numbers then*

$$x + y^{\text{I}} = (x + y)^{\text{I}}.$$

In order to illustrate the use of these axioms, we may prove that $1 + 3 = 4$, where, as expected,

(1)	$1 = 0^{\text{I}}$
(2)	$2 = 1^{\text{I}}$
(3)	$3 = 2^{\text{I}}$
(4)	$4 = 3^{\text{I}}$

We then have:

$$
\begin{aligned}
1 + 3 &= 1 + 2^{\text{I}} && \text{by (3)} \\
&= (1 + 2)^{\text{I}} && \text{by P7} \\
&= (1 + 1^{\text{I}})^{\text{I}} && \text{by (2)} \\
&= (1 + 1)^{\text{II}} && \text{by P7} \\
&= (1 + 0^{\text{I}})^{\text{II}} && \text{by (1)} \\
&= (1 + 0)^{\text{III}} && \text{by P7} \\
&= 1^{\text{III}} && \text{by P6} \\
&= 2^{\text{II}} && \text{by (2)} \\
&= 3^{\text{I}} && \text{by (3)} \\
&= 4 && \text{by (4)}
\end{aligned}
$$

A pair of postulates like P6 and P7 is said to provide an *inductive* or *recursive* "definition" of addition. From the standpoint of the theory of definition as formulated in §2.1, such "definitions" are not proper ones at all. In particular, they violate the criterion of eliminability in a much more profound way than do conditional definitions. For example, given P1–P7 we cannot eliminate the addition symbol from the arithmetical theorem:

$$x + y = y + x.$$

Recursive definitions, however, are close to being proper ones, and what we want to show is that given the additional apparatus of set theory, we can replace any recursive definition by an explicit, proper definition. Naturally if we were interested in considering only addition and perhaps multiplication and exponentiation, it would be possible to avoid the general theory of recursive definitions and justify each by special argument.* The advantage of the general theory is that it provides a clear picture of exactly what further means of definition are added to elementary number theory by the use of general set theory.

*This course is followed for example in Landau's classic little work [1930].

On the basis of the axioms introduced so far we can introduce the theory of recursive definitions by means of a theorem schema. Because we cannot prove the existence of the set of natural numbers or of any other infinite set, we cannot define the standard binary operations of arithmetic as proper set-theoretical functions. For the development of elementary number theory this is not a matter of great concern, but both for the theory of denumerable sets and for the theory of the real numbers in the next chapter, the existence of the set of natural numbers is essential. And granted existence of this set, existence of the binary operations on the natural numbers as functions is easily established.

Thus to have available from the beginning direct construction of arithmetical operations as certain sets, we introduce at this point the *axiom of infinity.*

$$(\exists A)[0 \in A \ \& \ (\forall B)(B \in A \to B \cup \{B\} \in A)].$$

The attempt to prove the existence of an infinite set of objects has a rather bizarre and sometimes tortured history. Proposition No. 66 of Dedekind's famous *Was sind und was sollen die Zahlen?*, first published in 1888, asserts that there is an infinite system. (Dedekind's systems correspond to our sets.) His proof is such a beautiful combination of mathematical reasoning and vague epistemology that I give a free translation of it here:

> PROOF. My world of ideas, i.e., the totality S of all things which can be objects of my thought, is infinite. For if s is an element of S, then the idea s', that s can be an object of my thought, is itself an element of S. If one considers the latter the image $\varphi(s)$ of the element s, then the hereby defined mapping φ of S has the property that the image S' is a part of S; and indeed S' is a proper part of S, because there are in S elements (e.g., my individual ego) which are different from every such idea s' and hence are not contained in S'. Finally, it is evident that if a and b are different elements of S, then their images a' and b' are also different; consequently the mapping φ is 1–1. Therefore S is infinite.*

Dedekind's proof depends, of course, on using his definition of infinite systems (or sets), but what is interesting is his excursion into the epistemology of ideas.† In no sense does this proof satisfy modern canons. Subtler arguments are to be found in Russell [1903, §339], but Russell later conceded that his arguments too were fallacious.††

---

*A similar argument is to be found in Bolzano [1851, §13].

†Epistemological criticisms of Dedekind's view are to be found in Russell [1920, pp. 139–140].

††Russell boldly begins §339: "That there are infinite classes is so evident that it will scarcely be denied. Since, however, it is capable of formal proof, it may be as well to prove it."

As far as I know the first unequivocally clear recognition that such an axiom is needed is to be found in Zermelo's epoch-making paper of 1908.* Zermelo's formulation is essentially the following:

$$(\exists A)\{0 \in A \ \& \ (\forall B)(B \in A \rightarrow \{B\} \in A)\}.$$

He constructed the natural numbers as $0, \{0\}, \{\{0\}\}, \{\{\{0\}\}\}, \dots$, which approach does not as easily generalize to the construction of infinite ordinals as does the one adopted in this book. A few years after Zermelo, Whitehead and Russell postulated an axiom of infinity in *Principia Mathematica*, and there is an interesting discussion in Ramsey [1926] of whether or not their axiom of infinity may be regarded as a logical truth.

Turning back to systematic considerations, we first define the set ω of all natural numbers and then leave as an exercise proof of the theorem that ω is non-empty. The axiom of infinity is essential for the proof.

DEFINITION 8. $\omega = \{A : A \text{ is a natural number}\}$.

We then have:

THEOREM 23. $A \in \omega$ if and only if A is a natural number.

It will be convenient for the sequel to use ω to give a "set" formulation of the principle of induction.

THEOREM 24.   *If*

    (i) $0 \in A$,
    (ii) *for every n if $n \in A$ then $n^1 \in A$,*
    then $\omega \subseteq A$.

We turn now to our main task of justifying definition by recursion. For functions of one argument, we want a theorem something like:

For every object x and every set G there is a unique F such that

    (i) *F is a function on ω,*
    (ii) *$F(0) = x$,*
    (iii) *for every n, $F(n^1) = G(F(n))$.*

It is important to notice that x need not be a natural number; it may be any individual or set (we use 'object' as a neutral term), and G need not be a function, nor if it is a function need it include $F(n)$ in its domain. Of course, when G is defective in one of these respects, then

$$G(F(n)) = 0,$$

by virtue of Definition 40 of §3.4.

---

*Moreover, Zermelo [1909] was the first to recognize that elementary number theory could be developed without the axiom of infinity.

For functions F of two arguments we want:

For any sets G and H there is a unique F such that for every m and n

(i) F is a function on $\omega \times \omega$,

(ii) $F(\langle m, 0\rangle) = H(m)$,

(iii) $F(\langle m, n'\rangle) = G(\langle m, F(\langle m, n\rangle)\rangle)$.

Note again that G and H need not be functions, although it is natural in defining the standard operations by recursion to have G and H be simple functions mapping natural numbers or pairs of natural numbers into natural numbers. For example, if F is meant to be the operation of addition, we set

$$H(m) = m$$

since

$$m + 0 = m,$$

and

(1) $$G(\langle m, F(\langle m, n\rangle)\rangle) = F(\langle m, n\rangle)'.$$

However, (1) is not quite correct; because the successor symbol does not designate a function in our set-theoretical universe, we do not know directly that G is a proper set-theoretical function. But this difficulty is easily remedied by the technique of defining a fragment of the intuitive successor function, corresponding to what we did earlier in the case of identity and membership.

DEFINITION 9. $\mathfrak{S}_A = \{\langle B, B'\rangle : B \in A\}$.

The expected theorem holds:

THEOREM 25. $\langle B, B'\rangle \in \mathfrak{S}_A \leftrightarrow B \in A$.

For simplicity of notation in this section, we further define:

DEFINITION 10. $\mathfrak{S} = \mathfrak{S}_\omega$.

Thus $\mathfrak{S}(0) = 1$ and $\mathfrak{S}(1) = 2$,
where we define, as already indicated:

DEFINITION 11.

$$1 = \{0\},$$
$$2 = \{0, \{0\}\}.$$

And we replace (1) above by

$$G(\langle m, F(\langle m, n\rangle)\rangle) = \mathfrak{S}(F(\langle m, n\rangle)).$$

We also define at this point $n - 1$ and $n - 2$ for any finite ordinal.

DEFINITION 12.

(i)   *If $A \neq 0$ & A is a finite ordinal then*
$$A - 1 = B \text{ if and only if } B^{\mathsf{I}} = A$$

(ii) *If $A \neq 0$ & $A \neq 1$ & A is a finite ordinal then*
$$A - 2 = B \text{ if and only if } (B^{\mathsf{I}})^{\mathsf{I}} = A.$$

Rather than state separate theorems for functions of one argument, functions of two arguments, etc., we may assert a general theorem on recursion for functions of r arguments. In the theorem we use the notation

(2) $$\langle m_0, m_1, \ldots, m_{r-2}, n \rangle$$

for r-tuples, which we now consider. First we define:

DEFINITION 13.   *x is an r-tuple of A if and only if $x \in A^r$.*

In other words, an r-tuple of A is a function from the set of natural numbers less than r to the set A. To justify the notation '$\langle x, y \rangle$' for couples (i.e, 2-tuples) when this notation has already been used for ordered pairs, we have the following theorem, whose proof we leave as an exercise:

THEOREM 26.   *$A^2 \approx A \times A$ under the function g such that for any $f \in A^2$, $g(f) = \langle f(0), f(1) \rangle$.*

Because A^2 is equipollent to $A \times A$, we shall use the same notation for couples and ordered pairs, and this ambiguity will be the source of no difficulty. In fact, in intuitive developments of set theory it is common to "identify" the two. Without using the notation (2), our fundamental theorem on recursion can be given the following somewhat unintuitive formulation:

For any sets G and H and any $r > 0$ there is a unique F such that

(i)   *F is a function on ω^r,*
(ii) *for every f, if $f \in \omega^r$ and $f(r - 1) = 0$ then*
$$F(f) = H(f \mid r - 1),$$
(iii) *for every f and every n, if $f \in \omega^r$ and $f(r - 1) = n$ then*

$$F((f \mid r - 1) \cup \{\langle r - 1, n^{\mathsf{I}} \rangle\}) = G((f \mid r - 1) \cup \{\langle r - 1, F(f) \rangle\}).$$

To reformulate the theorem in a schematic fashion after the manner of (2), we need:

DEFINITION SCHEMA 14.   *If $x_0, \ldots, x_{r-1} \in A$ then $\langle x_0, \ldots, x_{r-1} \rangle = f$ if and only if f is an r-tuple of A & $f(0) = x_0$ & $\ldots$ & $f(r - 1) = x_{r-1}$.*

Also, we modify in a standard manner the notation $f(x)$.

DEFINITION 15.

$$f_x = f(x).$$

Mainly, we want to write: f_0, f_1, etc. Finally, to formulate our theorem in familiar notation, we use 'm' in place of 'f' and when we write 'm_0', 'm_1' or the like we mean the value of the function m for the argument 0, etc., and not the numerical variable 'm_0', the numerical variable 'm_1', etc. This new use of the variable 'm' as a functional variable should not be the source of difficulty, for intuitively we may indeed treat 'm_0', 'm_1', etc. like numerical variables.

THEOREM 27. *For any sets G and H and any $r > 0$ there is a unique F such that for every m in ω^r,*

(i) *F is a function on ω^r,*

(ii) $F(\langle m_0, m_1, \ldots, m_{r-2}, 0 \rangle) = H(\langle m_0, m_1, \ldots, m_{r-2} \rangle)$,

(iii) *for every natural number n*

$$F(\langle m_0, m_1, \ldots, m_{r-2}, n' \rangle) = G(\langle m_0, m_1, \ldots, m_{r-2}, F(\langle m_0, \ldots, m_{r-2}, n \rangle) \rangle).$$

PROOF. For notational brevity we give the proof for the special case of $r = 2$, but the argument for arbitrary $r > 0$ is nearly identical and may be left as an exercise. We consider functions which for $r = 2$ could be defined on $\omega \times n'$ for some n. To parallel the argument for arbitrary r, we define:

$$\mathfrak{D}(n) = \{g : g \text{ is a function } \& \, \mathfrak{D}g = \{0, 1\} \, \& \, g(0) \in \omega \, \& \, g(1) \in n'\}.$$

First, by virtue of the axiom schema of separation there is a set A such that $f \in A$ if and only if:

(1) $f \in \mathcal{P}(\omega^2 \times (\mathcal{R}(G \cup H) \cup \{0\}))$,

and there is an n such that for every m

(2) f is a function on $\mathfrak{D}(n)$,

(3) $f(m, 0) = H(m)$,

(4) for every $p < n$

$$f(\langle m, p' \rangle) = G(\langle m, f(\langle m, p \rangle) \rangle).$$

The inclusion of $\{0\}$ in the power set of (1) takes care of the case when G and H are not defined. It is clear that (1) is implied by (2)-(4). Furthermore we easily see that $A \neq 0$, for the function f on $\mathfrak{D}(0)$ such that for every $m \in \omega$

(5) $f(m, 0) = H(m)$

is in A. (As indicated in (3) and (5) for the remainder of the proof we drop the pointed brackets to designate couples, since no serious confusion of meaning can result.)

(6) if $f, g \in A$ then either $f \subseteq g$ or $g \subseteq f$.

Let $\mathfrak{D}(n)$ be the domain of f and let $\mathfrak{D}(n_1)$ be the domain of g. Then $n \cap n_1$ is either n or n_1. Assume, for definiteness, that it is n. Now suppose there is a natural number $p < n$ such that for some m

(7) $$f(m, p) \neq g(m, p).$$

Let p^* be the smallest such number; that is, p^* is the $\mathcal{E}\omega$-first element of

$$\{p: p < n \ \& \ (\exists m)(f(m, p) \neq g(m, p))\}.$$

Now $p^* \neq 0$ since for every m

$$f(m, 0) = g(m, 0) = H(m).$$

We then have:

(8) $$f(m, p^*) = G(m, f(m, p^*-1)) \neq G(m, g(m, p^*-1)) = g(m, p^*).$$

But, on the other hand, since p^* is the smallest natural number for which (7) holds, we have:

$$f(m, p^*-1) = g(m, p^*-1)$$

and *a fortiori*

$$G(m, f(m, p^*-1)) = G(m, g(m, p^*-1)),$$

which contradicts (8) and renders our supposition (7) false. Thereby (6) is established.

Now, we define:

(9) $$F = \cup A.$$

It follows at once from (6) that F is a function, for if $\langle x, y \rangle \in f \in A$ and $\langle x, z \rangle \in g \in A$, then both couples belong either to f or to g, and hence $y = z$. Furthermore, it follows from (5) and (9) that for every m

$$F(m, 0) = H(m),$$

satisfying (ii) of the theorem.

We now turn our attention to (iii). If $\langle m, n^\iota \rangle$ is in the domain of F, then for some f in A, $\langle m, n^\iota \rangle$ is a member of the domain of f, and hence,

$$f(m, n^\iota) = G(m, f(m, n)),$$

whence

$$F(m, n^\iota) = G(m, F(m, n)).$$

To show that ω^2 is the domain of F, suppose that p is the smallest natural number such that for some m, $\langle m, p \rangle$ is not in the domain of F. Then in view of (5), $p \neq 0$, and thus $\langle m, p - 1 \rangle$ is in $\mathfrak{D}F$, and $F \in A$, but then also

$$(10) \qquad F \cup \bigcup_{m \in \omega} \{ \langle \langle m, p \rangle, G(m, F(m, p - 1)) \rangle \} \in A.$$

We may conclude from (10) that $\langle m, p \rangle \in \mathfrak{D}F$, contrary to our supposition. We conclude that ω^2 is the domain of F. Finally, we leave the simple proof that F is unique as an exercise. Q.E.D.

With this fundamental general theorem proved concerning the existence of a unique function F satisfying the definiens of a recursive definition, we may define the standard arithmetical operations without individual justifying theorems. We simply pick the appropriate functions G and H.

We first define addition of natural numbers by the recursion already indicated. We use the same symbol '+' for cardinal and finite ordinal addition (and later for arbitrary ordinal addition), but in any given context it will always be clear which is meant. It may be noted that the symbol for finite ordinal addition designates a set-theoretical entity, in particular, a certain set of ordered pairs, the first member of each pair being a couple. In contrast, the symbol for cardinal or general ordinal addition does not designate any set.

DEFINITION 16. $+ = f$ *if and only if*:

(i) f *is a function on* ω^2,

(ii) *for every* m

$$f(\langle m, 0 \rangle) = m,$$

(iii) *for every* m *and* n

$$f(\langle m, n^\text{I} \rangle) = \mathfrak{S}(f(\langle m, n \rangle)).$$

To obtain the usual notation, we define:

DEFINITION 17. $m + n = p$ *if and only if* $\langle \langle m, n \rangle, p \rangle \in +$.

As immediate consequences of Definition 16 we have:

THEOREM 28.

(i) $m + 0 = m$,

(ii) $m + n^\text{I} = (m + n)^\text{I}$.

The familiar commutative and associative laws of addition are asserted by the next pair of theorems. Their proofs illustrate typical uses of mathematical induction (via Theorem 24). Note that the corresponding theorems for cardinal arithmetic did not need to be proved inductively.

THEOREM 29. $m + n = n + m$.

PROOF.   We need two preliminary results:

(I) $0 + n = n$

(II) $m^\textrm{I} + n = (m + n)^\textrm{I}$,

each of which we prove inductively, using Theorem 24. To prove (I), we define:

(1) $Z = \{n \colon 0 + n = n\}$,

and we want to show that $Z = \omega$, that is, every natural number belongs to Z. First, by virtue of Theorem 28

$$0 \in Z.$$

Next assume that $n \in Z$, that is,

(2) $0 + n = n$.

Then, we have:

$$0 + n^\textrm{I} = (0 + n)^\textrm{I} \qquad \text{by Theorem 28}$$
$$= n^\textrm{I} \qquad \text{by (2)}.$$

Hence if $n \in Z$ then $n^\textrm{I} \in Z$ and thus by virtue of Theorem 24

$$Z = \omega$$

(since it follows from (1) that $Z \subseteq \omega$ and from Theorem 24 that $\omega \subseteq Z$).
  Now to prove (II) we define:

$$A(m) = \{n \colon m^\textrm{I} + n = (m + n)^\textrm{I}\}.$$

By two applications of Theorem 28

$$m^\textrm{I} + 0 = m^\textrm{I} = (m + 0)^\textrm{I},$$

and thus

$$0 \in A(m).$$

Now, assume that $n \in A(m)$, that is,

(3) $m^\textrm{I} + n = (m + n)^\textrm{I}$.

Then,

$$m^\textrm{I} + n^\textrm{I} = (m^\textrm{I} + n)^\textrm{I} \qquad \text{by Theorem 28}$$
$$= ((m + n)^\textrm{I})^\textrm{I} \qquad \text{by (3)}$$
$$= (m + n^\textrm{I})^\textrm{I} \qquad \text{by Theorem 28},$$

and thus $n^{\shortmid} \in A(m)$ whenever $n \in A(m)$. Whence by Theorem 24

$$A(m) = \omega,$$

which establishes (II), and completes our preparation for the main business at hand.

We define:

$$B(n) = \{m: m + n = n + m\}.$$

First, we have:

$$0 + n = n \qquad \text{by (I)}$$
$$= n + 0 \qquad \text{by Theorem 28,}$$

and therefore

$$0 \in B(n).$$

Now assume $m \in B(n)$, that is,

(4) $$m + n = n + m.$$

Then

$$m^{\shortmid} + n = (m + n)^{\shortmid} \qquad \text{by (II)}$$
$$= (n + m)^{\shortmid} \qquad \text{by (4)}$$
$$= n + m^{\shortmid} \qquad \text{by Theorem 28,}$$

whence $m^{\shortmid} \in B(n)$ whenever $m \in B(n)$, and we conclude that

$$B(n) = \omega,$$

the desired result.   Q.E.D.

The proof of the associative law for addition is similar in structure.

THEOREM 30. $(m + n) + p = m + (n + p)$.

PROOF.   We define:

$$A(m, n) = \{p: (m + n) + p = m + (n + \text{p})\}.$$

Now

$$(m + n) + 0 = m + n \qquad \text{by Theorem 28}$$
$$= m + (n + 0) \qquad \text{by Theorem 28 again,}$$

and therefore $0 \in A(m, n)$.

Let p be in $A(m, n)$.   Then

(1) $$(m + n) + p = m + (n + p),$$

and we have:

$$(m + n) + p^| = ((m + n) + p)^| \qquad \text{by Theorem 28}$$
$$= (m + (n + p))^| \qquad \text{by (1)}$$
$$= m + (n + p)^| \qquad \text{by Theorem 28}$$
$$= m + (n + p^|) \qquad \text{by Theorem 28,}$$

whence $p^| \in A(m, n)$, and we may conclude:

$$A(m, n) = \omega. \qquad\qquad \text{Q.E.D.}$$

We state without proof a familiar theorem.

THEOREM 31. *If $m \leq n$ then there is a unique natural number p such that $m + p = n$.*

We now turn to the definition of multiplication. If we were proceeding without set theory, we would add to P6 and P7 two more axioms:

P8. *If x is a natural number then $x \cdot 0 = 0$.*

P9. *If x and y are natural numbers then $x \cdot y^| = (x \cdot y) + x$.*

These two axioms accurately forecast the definition we use.

DEFINITION 18. $\cdot = f$ *if and only if:*

(i) *f is a function on ω^2,*

(ii) *for every m*

$$f(\langle m, 0 \rangle) = 0,$$

(iii) *for every m and n*

$$f(\langle m, n^| \rangle) = f(\langle m, n \rangle) + m.$$

Like Definition 17 we have:

DEFINITION 19. $m \cdot n = p$ *if and only if $\langle \langle m, n \rangle, p \rangle \in \cdot$.*

When no confusion will result we designate multiplication by juxtaposition rather than the dot.

We leave proofs of the following three theorems as exercises.

THEOREM 32. $mn = nm.$

THEOREM 33. $m(n + p) = mn + mp.$

THEOREM 34. $(mn)p = m(np).$

A number of further theorems are stated as exercises.

We conclude this section with the definition of the exponential operation. The recursive scheme is:

$$m^0 = 1,$$
$$m^{n^!} = m^n \cdot m.$$

The appropriate formal definition is:

DEFINITION 20.   exp $= f$ *if and only if*:

(i) f *is a function on* ω^2,
(ii) *for every* m

$$f(\langle m, 0\rangle) = 1,$$

(iii) *for every* m *and* n

$$f(\langle m, n^!\rangle) = f(\langle m, n\rangle) \cdot m.$$

And in deference to orthodox notation:

DEFINITION 21. $m^n = \exp{(m, n)}$.

In the next chapter, which is concerned with the construction of the rational and real numbers, we shall assume that the elementary arithmetic of the integers has been completely developed, for with Theorem 27 at hand it is perfectly obvious how to continue developments. Actually a number of details are given in the exercises.

In Chapter 7 it will be indicated how Theorem 27 may be generalized to what is termed a course-of-values recursion. Thus, when F is a function of one argument we have:

(1) $$F(n) = G(F \mid n),$$

that is, the value of F for the argument n may depend on all preceding arguments and values of F, which is indicated on the right-hand side of (1) by the restriction of the domain of F to n. A simple example of such a course of values recursion is the definition of the appropriate function to yield the Fibonacci sequence

$$0, 1, 1, 2, 3, 5, 8, 13, \ldots$$

Intuitively, except for the first two terms, each term is simply the sum of the preceding two. The recursion scheme is:

$$F(0) = 0$$
$$F(1) = 1$$
$$F(n + 1) = F(n - 1) + F(n),$$

that is, the general term $F(n + 1)$ depends not only on $F(n)$ but also on $F(n - 1)$.

We conclude this section with the definition of ordinary finiteness already anticipated in §4.2 and we leave as an exercise the proof of the equivalence of this definition with that of Tarski, which was used in §4.2.

DEFINITION 22. *A set is ordinary finite if and only if it is equipollent to some natural number.*

THEOREM 35. *A set is ordinary finite if and only if it is finite in the sense of Tarski.*

We also have:

THEOREM 36. *A finite set is equipollent to exactly one natural number.*

On the basis of this theorem and the definition of natural numbers, it is not difficult to prove that Stäckel's [1907] definition of finiteness in terms of double well-ordering is equivalent to Tarski's.

THEOREM 37. *A set is finite in the sense of Tarski if and only if it can be doubly well-ordered, i.e., if and only if there is a relation R such that both R and $\breve{R}$ well-order the set.*

EXERCISES

1. Prove: *A is a natural number if and only if A is an ordinal, and for every B if $B \in A^{\text{l}}$ then there is an ordinal C such that $B = C^{\text{l}}$ or $B = 0$.* (This equivalence is sometimes used as the definition of natural numbers.)

2. Give a counterexample to: *A is a natural number if and only if A is an ordinal and either $A = 0$ or there is an ordinal B such that $A = B^{\text{l}}$.*

3. Prove Theorem 21.

4. Give a detailed proof of Theorem 23.

5. Give a counterexample to show that not every subset of a natural number is a natural number.

6. Prove that if A is a natural number and B is the $\mathcal{E}A$-last element of A then $A = B^{\text{l}}$.

7. Does the assertion of Exercise 6 hold for arbitrary ordinals?

8. Prove that if A is a natural number then $\bigcup (A^{\text{l}})$ is also.

9. If $A \subset \omega$ is it true that $\bigcup A$ is a natural number? If so, prove it. If not give a counterexample.

10. If A is an ordinal, what set is $\bigcap A$?

11. If $A \subseteq \omega$, is $\bigcap A$ a natural number?

12. Prove that $\bigcup \omega = \omega$.

13. Complete the proof of Theorem 27 by showing that F is unique.

14. Indicate what changes are required in the proof of Theorem 27 to make it adequate for arbitrary r.

15. Prove Theorem 31.

16. Prove Theorem 32.

17. Prove Theorems 33 and 34.

18. Give a recursive scheme for and then explicitly define:
  (a) the factorial operation $n!$;
  (b) the Fibonacci sequence mentioned at the end of the section (by slight reformulation of the recursion given for the sequence it may be brought within the scope of Theorem 27).
19. Prove the familiar elementary facts about the exponentiation operation for natural numbers.
20. Prove Exercises 4-12 of §4.4 for natural numbers.
21. Prove Theorem 35.
22. Prove Theorem 36.
23. Prove Theorem 37.

§ **5.3 Denumerable Sets.**   A denumerable set is one which is equipollent to the set of natural numbers. Such a set affords the simplest example of an infinite set. We begin with some general theorems about infinite sets.

DEFINITION 23.   *A set is infinite if and only if it is not finite.*

The first two theorems may be easily proved by using results in §4.2.

THEOREM 38.   *If A is infinite and $A \approx B$ then B is infinite.*

THEOREM 39.   *If $A \subseteq B$ and A is infinite then B is infinite.*

Of somewhat more interest is the following theorem which provides a necessary and sufficient condition for a set to be infinite.

THEOREM 40.   *A set A is infinite if and only if for every natural number n there is a subset of A equipollent to n.*

PROOF. [Necessity]. By virtue of the axiom schema of separation there is a set C such that for every n

$$n \in C \leftrightarrow n \in \omega \,\&\, (\exists B)(B \subseteq A \,\&\, B \approx n).$$

We show by induction that $C = \omega$.
    Since the empty set is a subset of every set,

$$0 \in C.$$

Now assume that $n \in C$. Then by hypothesis there is a subset B of A such that

$$B \approx n.$$

Because A is infinite, $B \neq A$, for if $A = B$ then $A \approx n$ and A would be finite (Theorem 35). Let x therefore be an element of $A \sim B$. Then

$$B \cup \{x\} \subseteq A$$

and clearly in view of Theorem 4 of §4.1

$$B \cup \{x\} \approx n'$$

whence $n' \in C$.

[Sufficiency]. Suppose if possible that A is finite. Then (Theorem 35), for some n

$$A \approx n.$$

But by hypothesis there must be a non-empty proper subset B of A such that $B \approx n'$. Let $x \in B$, then $B \sim \{x\} \approx n$, whence A is equipollent to one of its proper subsets, namely, $B \sim \{x\}$, and this contradicts Theorem 46 of §4.2.   Q.E.D.

Using this theorem and some earlier results it is easy to prove:

THEOREM 41. *The set ω of natural numbers is infinite.*

A more difficult theorem is the following, which expresses a necessary and sufficient condition for a set to be finite. The proof of sufficiency is the difficult part because it involves definition by recursion of a function (Theorem 27).

THEOREM 42. *A is finite if and only if $A \prec \omega$.*

PROOF. [Necessity]. This theorem follows immediately from the preceding theorem and Theorem 45 of §4.2.

[Sufficiency]. By hypothesis $A \prec \omega$. Let B be a proper subset of ω such that $A \approx B$. If B has a largest natural number, say m, then $B \subseteq m'$, and since m' is finite, so is B and consequently also A. The difficult thing to show is that indeed B must have a largest natural number. (We assume B is non-empty; otherwise, the proof is trivial.) We use an indirect proof. Suppose that B has no largest natural number, i.e., given any natural number there is always a larger one in B. Using Theorem 27 we define the function F on ω as follows:

(1) $F(0) = \varepsilon\omega$-first element of B.

(2) $F(n') = \varepsilon\omega$-first element of $B \sim F(n)'$.

(Note that it is simple to define functions G and H corresponding to the right-hand sides of (1) and (2).)
Since B has no largest element, it is clear that for every n and m if $n \neq m$ then

$$F(n) \neq F(m),$$

whence $B \approx \omega$ under F, contrary to our hypothesis, and our supposition is false.   Q.E.D.

We now turn to denumerable sets and we begin by restating the definition given at the beginning of the section.

DEFINITION 24. *A set is denumerable if and only if it is equipollent to the set ω of all natural numbers.*

As an immediate consequence of Theorems 38 and 41 we have:

THEOREM 43. *Every denumerable set is infinite.*

Somewhat surprisingly, every known proof of the partial converse, namely that every infinite set has a denumerable subset, requires the axiom of choice. In fact, this converse is the essential step in showing that ordinary infinity implies Dedekind infinity, which concept we now formally define. The two theorems following the definition are immediate consequences of earlier results.

DEFINITION 25. *A set is Dedekind infinite if and only if it is not Dedekind finite.*

THEOREM 44. *A set is Dedekind infinite if and only if it has a proper subset equipollent to it.*

THEOREM 45. *A is Dedekind infinite if and only if $A \approx A \cup \{A\}$.*

The contrapositive of Theorem 46 of Chapter 4 is:

THEOREM 46. *If a set is Dedekind infinite then it is infinite.*

We next turn to two more difficult theorems which together establish another necessary and sufficient condition for a set to be Dedekind infinite.

THEOREM 47. *Every set which has a denumerable subset is Dedekind infinite.*

PROOF. Let B be a denumerable subset of A. Then

$$B \approx \omega$$

under some 1–1 function, say f. Now consider the function g defined on A by

$$g(x) = \begin{cases} \breve{f}(f(x)\prime) & \text{if } x \in \mathfrak{D}f \\ x & \text{if } x \in A \sim \mathfrak{D}f. \end{cases}$$

(Intuitively if $x \in \mathfrak{D}f$ and $f(x) = n$, write: x_n. Then if $x \in \mathfrak{D}f$ we simply have: $g(x_n) = x_{n+1}$.) Obviously

$$g \text{ is 1–1}$$
$$\mathfrak{D}g = A$$
$$\mathfrak{R}g = A \sim \{\breve{f}(0)\},$$

whence

$$A \approx A \sim \{\breve{f}(0)\},$$

establishing our theorem. Q.E.D.

We now prove the converse, which requires recursive definition of a suitable function.

Theorem 48. *Every Dedekind infinite set has a denumerable subset.*

proof. Let A be a Dedekind infinite set and let B be a proper subset of A such that $A \approx B$ under the function f, say. Let x^* be an element in $A \sim B$. We define recursively (Theorem 27) a function g on ω:

$$g(0) = x^*$$
$$g(n') = f(g(n)).$$

Clearly, for each n, $g(n) \in A$. What we need to prove is that the function g is 1–1. Suppose, by way of contradiction, that g is not 1–1. Then let p be the smallest natural number such that for some $q < p$

(1) $$g(p) = g(q).$$

Obviously $p \neq 0$. Moreover, since

(2) $$g(p) = f(g(p - 1)),$$

we know that $g(p) \in B$, and since $g(0) = x^* \not\in B$, we have

$$g(p) \neq g(0)$$

and thus from (1)

$$g(q) \neq g(0),$$

whence

$$q \neq 0.$$

Thus

(3) $$g(q) = f(g(q - 1)),$$

and from (1)–(3) it follows that

$$f(g(p - 1)) = f(g(q - 1)),$$

and since f is 1–1

$$g(p - 1) = g(q - 1),$$

which contradicts our supposition that p is the smallest natural number satisfying (1) with $q < p$. Q.E.D.

Theorem 49. *Every subset of a denumerable set is either denumerable or finite.*

proof. Let A be a denumerable set and let $B \subseteq A$. Then $B \preceq A$. If $B \approx A$ then B is denumerable. If $B \prec A$, then $B \prec \omega$, since $A \approx \omega$, and thus by Theorem 42, B is finite. Q.E.D.

This theorem represents a significant application of Theorem 42. The simplicity of the proof depends wholly on prior proof of Theorem 42.

THEOREM 50. *If A is finite and B is denumerable then $A \cup B$ is denumerable.*

PROOF. Let

$$C = A \sim B;$$

from the hypothesis of the theorem C is finite, and thus there is a natural number n such that (Theorem 35)

$$C \approx n$$

under f, say. Now B is denumerable, whence

$$B \approx \omega$$

under a function g, say. We now define a new function h on $A \cup B$:

$$h(x) = \begin{cases} f(x) & \text{if } x \in C \\ g(x) + n & \text{if } x \in B. \end{cases}$$

Clearly h is 1–1 and it maps $A \cup B$ onto ω. Q.E.D.

The proof, which we omit, of the next theorem is more difficult.

THEOREM 51. *If A and B are denumerable then $A \cup B$ is denumerable.*

We have some corresponding theorems about the denumerability of Cartesian products. In this case we prove the second one.

THEOREM 52. *If A is finite, but non-empty, and B is denumerable then $A \times B$ is denumerable.*

THEOREM 53. $\omega \times \omega \approx \omega.$

PROOF. The proof rests upon arranging the double-sequence

$$\langle 0, 0 \rangle, \langle 0, 1 \rangle, \langle 0, 2 \rangle, \ldots$$
$$\langle 1, 0 \rangle, \langle 1, 1 \rangle, \langle 1, 2 \rangle, \ldots$$
$$\langle 2, 0 \rangle, \langle 2, 1 \rangle, \langle 2, 2 \rangle, \ldots$$

into a single sequence by the method of diagonals (we proceed along diagonals):

(1) $\langle 0, 0 \rangle, \langle 0, 1 \rangle, \langle 1, 0 \rangle, \langle 0, 2 \rangle, \langle 1, 1 \rangle, \langle 2, 0 \rangle, \ldots .$

Another way of characterizing (1) is to say that we order first according to the sum of the two numbers, and then by first member within a fixed sum. We define on $\omega \times \omega$ a function f such that

$$f(\langle 0, 0 \rangle) = 0$$

$$f(\langle 0, 1 \rangle) = 1$$

$$f(\langle 1, 0 \rangle) = 2$$

$$f(\langle 0, 2 \rangle) = 3$$

. . . .

It is clear that f maps $\omega \times \omega$ onto ω in 1–1 fashion; finding the exact arithmetical form of f we leave as an exercise.   Q.E.D.

THEOREM 54.   *If $n \neq 0$ then $\omega^n \approx \omega$.*

THEOREM 55.   *If A and B are denumerable then $A \times B$ is denumerable.*

THEOREM 56.   *If A is denumerable and $n \neq 0$ then A^n is denumerable.*

We now turn to some theorems about infinite and transfinite cardinals. Relevant definitions and proofs of the theorems all depend on the special axiom for cardinal numbers. It should also be clear from earlier remarks that every known proof that a cardinal is transfinite if and only if it is infinite depends on the axiom of choice.

†DEFINITION 26.   *x is an infinite cardinal if and only if there is an infinite set A such that $\mathcal{K}(A) = x$.*

†DEFINITION 27.   *x is a transfinite cardinal if and only if there is a Dedekind infinite set A such that $\mathcal{K}(A) = x$.*

On the basis of theorems previously proved, the following four results are easily derived.

†THEOREM 57.   *$\mathfrak{n}$ is an infinite cardinal if and only if for every finite cardinal $\mathfrak{m}$, $\mathfrak{m} < \mathfrak{n}$.*

†THEOREM 58.   *$\mathfrak{n}$ is a transfinite cardinal if and only if $\mathfrak{n} = \mathfrak{n} + 1$.*

†THEOREM 59.   *If a cardinal is a transfinite cardinal then it is an infinite cardinal.*

†THEOREM 60.   *If $\mathfrak{m}$ is a transfinite or infinite cardinal and $\mathfrak{m} \leq \mathfrak{n}$ then $\mathfrak{n}$ is a transfinite or infinite cardinal respectively.*

A theorem which takes a certain amount of proving (we leave the proof as an exercise) is that the sum or product of a transfinite or infinite cardinal with any other cardinal (excepting 0 for product) is also a transfinite or infinite cardinal.

†Theorem 61. *If* m *is a transfinite or infinite cardinal then*

   (i) $m + n$ *is a transfinite or infinite cardinal respectively,*

   (ii) mn *and* m^n *are transfinite or infinite cardinals, respectively, provided* $n \neq 0$;

   (iii) n^m *is a transfinite or infinite cardinal respectively, provided* $1 < n$.

We may use some of these results, particularly Theorem 58, to prove an inequality for transfinite numbers, which does not hold for arbitrary cardinals. Note that the method of proof is specific to transfinite cardinals and will not work for infinite cardinals (without the axiom of choice).

†Theorem 62. *If* m *and* n *are transfinite cardinals, then*

$$m + n \leq mn.$$

proof.   On the basis of Theorem 58

$$m = m + 1$$
$$n = n + 1,$$

whence using the distributive and commutative laws of §4.3

(1) $mn = (m + 1)(n + 1) = mn + m + n + 1 = (m + n) + (mn + 1).$

Primarily on the basis of Theorem 74 of §4.3, we know that

$$m \leq m + n$$

and

$$n \leq mn + 1,$$

whence, again using Theorem 74,

(2) $m + n \leq (m + n) + (mn + 1).$

Our theorem follows from (1) and (2).   Q.E.D.

By use of the axiom of choice the inequality in the theorem just proved may be strengthened to an equality, that is, on the basis of assuming the axiom of choice, we may prove that the sum of two transfinite cardinals is equal to their product. Note that the theorem is false for finite cardinals, since $n \cdot 1 < n + 1$.

To convert some of the theorems about denumerable sets into theorems about cardinal numbers, we need to define the cardinal $\aleph_0$ of denumerable sets. ($\aleph_0$ is read 'aleph null'; the letter '$\aleph$' is the first letter of the Hebrew alphabet. This notation originates with Cantor.)

†Definition 28. $\aleph_0 = \mathcal{K}(\omega)$.

As immediate consequences of Theorems 50 and 51, and the definition of cardinal addition in §4.3, we have:

†Theorem 63.

(i)   *If n is a finite cardinal then $\aleph_0 + n = \aleph_0$,*
(ii) $\aleph_0 + \aleph_0 = \aleph_0$.

Proof of the following useful theorem we leave as an exercise.

†Theorem 64.   *The following three conditions are equivalent*:

(i) m *is a transfinite cardinal*,
(ii) $\aleph_0 \leq m$,
(iii)   *there is a cardinal n such that $\aleph_0 + n = m$.*

On the basis of Theorems 52 and 53, we have at once:

†Theorem 65.

(i)   *If n is a finite cardinal and $n \neq 0$ then $n\aleph_0 = \aleph_0$,*
(ii) $\aleph_0\aleph_0 = \aleph_0$.

And on the basis of Theorem 54:

†Theorem 66.   *If n is a finite cardinal and $n \neq 0$ then $\aleph_0{}^n = \aleph_0$.*

We now prove two further facts about transfinite cardinals.

†Theorem 67.   *If u is a transfinite cardinal then*

(i) $u + \aleph_0 = u$,
(ii)   *if n is a finite cardinal $u + n = u$.*

PROOF.   By virtue of Theorem 64 there is an n such that

$$\aleph_0 + n = u,$$

whence

$$u + \aleph_0 = (\aleph_0 + n) + \aleph_0$$

$$= n + (\aleph_0 + \aleph_0)$$

$$= n + \aleph_0 \qquad \text{by Theorem 63}$$

$$= \aleph_0 + n$$

$$= u,$$

which proves (i). Similar methods establish (ii):

$$u + n = (u + \aleph_0) + n \qquad \text{by (i)}$$

$$= u + (\aleph_0 + n)$$

$$= u + \aleph_0 \qquad \text{by Theorem 63}$$

$$= u \qquad \text{by (i) again.   Q.E.D.}$$

1. Prove Theorems 38 and 39.
2. Prove Theorem 41.
3. Prove Theorem 51.
4. Prove Theorem 52.
5. Find the exact arithmetical form of the function f in the proof of Theorem 53.
6. Prove Theorem 54.
7. Prove Theorems 55 and 56.
8. Prove that if the domain of a function is denumerable the function is denumerable and its range is either finite or denumerable.
9. Prove that if A is denumerable then there is a family B of sets such that
    (i) B is denumerable,
    (ii)   if $C \in B$ then C is denumerable,
    (iii) if $C, D \in B$ and $C \neq D$ then $C \cap D = 0$,
    (iv) $\cup B = A$.
(Known proofs of the converse of this theorem require the axiom of choice.)
10. Prove Theorems 57-60.
11. Prove Theorem 61.
12. Prove Theorem 64.
13. Prove that if $\mathfrak{n}$ is a finite cardinal then $\mathfrak{n} < \aleph_0$.

CHAPTER 6

RATIONAL NUMBERS AND REAL NUMBERS*

§ 6.1 Introduction. To show that our axioms for set theory are adequate to permit the systematic development of classical mathematics, it is not sufficient merely to construct the natural numbers as we did in the previous chapter. At the very least we need to show that we can construct entities which have all the expected properties of the real numbers.

The two basic set-theoretical methods of constructing the real numbers out of the natural numbers are due to Cantor and Dedekind, but Bertrand Russell also deserves credit for making clear the exact character of these constructions and for being completely explicit about identifying the real numbers as the constructed entities.

Antecedent to the construction of the real numbers is the construction of the rational numbers (intuitively a rational member is a number which can be represented as the ratio of two integers). Several alternative courses of development can be followed:

I	II
Natural numbers	Natural numbers
Integers	Non-negative fractions
Fractions	Non-negative rational numbers
Rational numbers	Non-negative real numbers
Real numbers	All real numbers

III

Natural numbers
Non-negative fractions
Non-negative rational numbers
All rational numbers
Real numbers

*This chapter may be omitted without loss of continuity.

Several variants of these three courses are possible depending upon the choice of a level at which to introduce negative numbers. Course III will be adopted here. Non-negative fractions are defined as ordered pairs of non-negative integers. Thus the fraction $\frac{1}{2} = \langle 1, 2 \rangle$. Then non-negative rational numbers are defined as certain equivalence classes of fractions. For instance, the non-negative rational number $[\frac{1}{2}]$ corresponding to the fraction $\frac{1}{2}$ is the set of all fractions m/n such that $n = 2m$. To get to all rational numbers, we go up another level of abstraction. We say that two ordered pairs $\langle x, y \rangle$ and $\langle u, v \rangle$ of non-negative rational numbers are equivalent when

$$x + v = y + u,$$

and a rational number is just an equivalence class of such ordered pairs. Perhaps it may seem odd to distinguish between the fraction $\frac{1}{2}$, the non-negative rational number $[\frac{1}{2}]$, the ordered pair $\langle [\frac{1}{2}], [\frac{0}{1}] \rangle$, and the rational number $[\langle [\frac{1}{2}], [\frac{0}{1}] \rangle]$. But as we shall see, each level of abstraction is built on the preceding one in a natural way.

With all the rational numbers available, the Dedekind procedure for constructing the real numbers is to define a *section* or *cut* of the rationals as an ordered pair $\langle A, B \rangle$ of sets such that

  (i) A and B are both non-empty,
  (ii) $A \cup B =$ the set of rationals,
  (iii) if $x \in A$ and $y \in B$ then $x < y$.

A is called the *lower class* and B the *upper class*, since every element of A precedes every element of B. A *real number* is then simply a section of the rationals.

A somewhat simpler definition was given by Peano and Russell. As suggested by a cursory inspection of the above definition of sections, why not dispense with either the set A or the set B? We are led to: a *lower cut* or *lower segment* of the rationals is a set A such that

  (i) $A \neq 0$,
  (ii) $A \subset \mathrm{Ra}$,
  (iii) if $x \in A$ and $y \in \mathrm{Ra} \sim A$ then $x < y$,
  (iv)   for every x in A there is a y in A such that $x < y$, where Ra
          is the set of rational numbers.

The proof that lower segments of the rationals have all the expected properties of the real numbers is given in great detail in Landau [1930].

Cantor's approach to the real numbers is less algebraic and more analytic in character. He uses the basic notion of a *sequence* of rationals, that is, the notion of a function whose domain is ω and whose range is a subset of

Ra. A sequence x is a *Cauchy sequence* if for every positive rational number ϵ there is a natural number N such that for all $m, n > N$

$$|x_n - x_m| < \epsilon.$$

(Using traditional notation, we write x_n, instead of $x(n)$.)*
Following Cantor, two Cauchy sequences x and y are said to be *equivalent* if for every positive rational number ϵ there is a natural number N such that for all $n > N$

$$|x_n - y_n| < \epsilon.$$

Real numbers are then defined as equivalence classes of Cauchy sequences.
   In this chapter we follow Cantor rather than Dedekind for the reason that when Cauchy sequences are used the methods of proof are much more characteristic of general methods in analysis, in particular of those employed in the theory of infinite series.

§ 6.2 Fractions.

Turning now to systematic developments we begin by constructing the non-negative fractions, which for brevity we shall simply call fractions. As in the previous chapter, we shall use the variables 'm', 'n', 'p', 'q', and 'r', with or without subscripts, for natural numbers.

   DEFINITION 1.   *x is a fraction* $\leftrightarrow (\exists m)(\exists n)(n \neq 0 \ \& \ x = \langle m, n \rangle)$.

Intuitively, $x = \dfrac{m}{n}$, and to have available this standard notation, we define:

   DEFINITION 2.   *If $n \neq 0$ then* $\dfrac{m}{n} = \langle m, n \rangle$.

And we shall also find it convenient to have the set **Fr** of fractions at hand.

   DEFINITION 3.   **Fr** $= \{x \colon x \text{ is a fraction}\}$.

In all definitional uses of the abstraction notation in this chapter it will be obvious that the set defined is the intuitively appropriate one and not the empty set. Thus we shall neither state nor prove theorems like:

$$x \in \mathbf{Fr} \leftrightarrow x \text{ is a fraction.}$$

We now define the relation $\simeq_f$ (the subscript is for 'fraction') such that if $n_1 \neq 0$ and $n_2 \neq 0$ then

$$\frac{m_1}{n_1} \simeq_f \frac{m_2}{n_2} \leftrightarrow m_1 n_2 = m_2 n_1.$$

---

*The name *Cauchy sequence* honors the great French mathematician A. L. Cauchy (1789–1857).

DEFINITION 4. $\simeq_f = \left\{ \langle \dfrac{m_1}{n_1}, \dfrac{m_2}{n_2} \rangle : n_1 \neq 0 \ \& \ n_2 \neq 0 \ \& \right.$

$$\left. m_1 n_2 = m_2 n_1 \right\}.$$

As our first theorem we then have:

THEOREM 1. $\simeq_f$ *is an equivalence relation on* **Fr**.

PROOF.   To indicate how the proof goes, we prove that $\simeq_f$ is reflexive in **Fr**. Let m/n be any fraction, then $mn = mn$, whence $m/n \simeq_f m/n$. Q.E.D.

In all these proofs elementary facts about operations and relations on the natural numbers are used without explicit reference.

THEOREM 2.   *If* $\dfrac{m}{n} \in$ **Fr** *and* $p \neq 0$ *then* $\dfrac{m}{n} \simeq_f \dfrac{mp}{np}$.

We now define *less than* for fractions.

DEFINITION 5. $<_f = \left\{ \langle \dfrac{m_1}{n_1}, \dfrac{m_2}{n_2} \rangle : n_1 \neq 0 \ \& \ n_2 \neq 0 \ \& \right.$

$$\left. m_1 n_2 < m_2 n_1 \right\}.$$

Two expected theorems are:

THEOREM 3. $<_f$ *is a strict partial ordering of* **Fr**.

THEOREM 4.   *If* $x,y \in$ **Fr** *then exactly one of the following:* $x \simeq_f y$, $x <_f y, \ y <_f x$.

Since $<_f$ is a set, a very simple definition of *greater than* for fractions is possible: it is the converse of less than.

DEFINITION 6. $>_f = \breve{<}_f$.

The next two theorems state, respectively, that there is no greatest fraction, and that between any two fractions there is a third.

THEOREM 5.   *If* $x \in$ **Fr** *then there is a* $y \in$ **Fr** *such that* $x <_f y$.

THEOREM 6.   *If* $x, y \in$ **Fr** *and* $x <_f y$ *then there is a* $z \in$ **Fr** *such that*

$$x <_f z <_f y.$$

The last theorem expresses an important property that relations may or may not have in a set. If the field of R contains A and for any x,y in A, if xRy there is a z in A such that

$$x \ R \ z \ \& \ z \ R \ y,$$

then R is said to be *dense* in A. Thus the theorem says that $<_f$ is dense in **Fr**.

The next theorem says that the relation $\simeq_f$ has an expected substitution property with respect to less than for fractions.

THEOREM 7. *If $x,y,u,v \in$ Fr & $x <_f y$ & $x \simeq_f u$ & $y \simeq_f v$ then $u <_f v$.*

We now define *addition* of fractions.

DEFINITION 7. $F+ = \left\{ \left\langle \dfrac{m_1}{n_1}, \dfrac{m_2}{n_2}, \dfrac{m_3}{n_3} \right\rangle : n_1 \neq 0 \text{ & } n_2 \neq 0 \text{ & } n_3 \neq 0 \right.$

$$\left. \text{ & } \frac{m_1 n_2 + n_1 m_2}{n_1 n_2} = \frac{m_3}{n_3} \right\}.$$

To justify the appropriate notation, we have:

THEOREM 8. *If $x,y \in$ Fr then there is a unique z in Fr such that $\langle x, y, z \rangle \in F+$.*

DEFINITION 8. *If $x,y,z \in$ Fr then $x + y = z \leftrightarrow \langle x, y, z \rangle \in F+$.*

Note that since the previous definition of the symbol '$+$' for ordinals was conditional, no confusion can arise from the omission of the subscript here. Obviously this is not the case for less than. By a similar procedure the subscript could have been omitted from '$\simeq_f$'.

We have the expected battery of theorems. The first one states that equivalence of fractions has the substitution property with respect to addition of fractions. This theorem, like Theorem 7, is crucial for developing the theory of rational numbers in the next section.

THEOREM 9. *If $x,y,u,v \in$ Fr & $x \simeq_f u$ & $y \simeq_f v$ then $x + y \simeq_f u + v$.*

THEOREM 10. *If $n \neq 0$ then $\dfrac{m_1}{n} + \dfrac{m_2}{n} \simeq_f \dfrac{m_1 + m_2}{n}$.*

THEOREM 11. *Addition of fractions is commutative and associative.*

PROOF. We only prove associativity.

$$\left(\frac{m_1}{n_1} + \frac{m_2}{n_2} \right) + \frac{m_3}{n_3} = \left(\frac{m_1 n_2 + n_1 m_2}{n_1 n_2} \right) + \frac{m_3}{n_3}$$

$$= \frac{(m_1 n_2 + n_1 m_2)n_3 + (n_1 n_2)m_3}{(n_1 n_2)n_3}$$

$$= \frac{m_1 n_2 n_3 + n_1 m_2 n_3 + n_1 n_2 m_3}{n_1 n_2 n_3}$$

$$= \frac{m_1(n_2 n_3) + n_1(m_2 n_3 + n_2 m_3)}{n_1(n_2 n_3)}$$

$$= \frac{m_1}{n_1} + \left(\frac{m_2 n_3 + n_2 m_3}{n_2 n_3} \right)$$

$$= \frac{m_1}{n_1} + \left(\frac{m_2}{n_2} + \frac{m_3}{n_3} \right). \qquad \text{Q.E.D.}$$

THEOREM 12. *Addition of fractions is monotonic with respect to the relation less than for fractions, that is, if x, y, $z \in$ Fr and $x <_f y$ then*

$$x + z <_f y + z$$

and

$$z + x <_f z + y.$$

THEOREM 13. *Addition of fractions has the cancellation property with respect to less than for fractions, that is, if x, y, $z \in$ Fr and*

either $x + z <_f y + z$

or $z + x <_f z + y$

then

$$x <_f y.$$

When we say that an operation simply has the cancellation property, we mean with respect to the identity relation. Thus a binary operation $\star$ has the *cancellation property* if and only if whenever $x \star z = y \star z$ or $z \star x = z \star y$ then $x = y$. We leave as an exercise the problem of determining if addition and multiplication of fractions have the cancellation property (with respect to identity).

We now define multiplication of fractions.

DEFINITION 9. $F^{\cdot} = \left\{ \left\langle \dfrac{m_1}{n_1}, \dfrac{m_2}{n_2}, \dfrac{m_3}{n_3} \right\rangle : n_1 \neq 0 \,\&\, n_2 \neq 0 \,\&\, n_3 \neq 0 \right.$

$$\left. \&\, \frac{m_1 m_2}{n_1 n_2} = \frac{m_3}{n_3} \right\}.$$

THEOREM 14. *If $x, y \in$ Fr then there is a unique z in Fr such that $\langle x, y, z \rangle \in F^{\cdot}$.*

DEFINITION 10. *If $x, y, z \in$ Fr then $xy = z \leftrightarrow \langle x, y, z \rangle \in F^{\cdot}$.*

THEOREM 15. *If $x, y, u, v \in$ Fr $\&$ $x \simeq_f u$ $\&$ $y \simeq_f v$ then $xy \simeq_f uv$.*

THEOREM 16. *Multiplication of fractions is commutative and associative; moreover, it is distributive (both from the left and right) with respect to addition of fractions.*

THEOREM 17. *Multiplication of fractions has the cancellation property with respect to less than for fractions.*

The next theorem essentially says that division of fractions except by zero is always possible.

THEOREM 18. *If $x, y \in$ Fr and $y >_f \frac{0}{1}$ then there is a z in Fr such that $x \simeq_f yz$.*

PROOF. Let

$$x = \frac{m_1}{n_1}$$

$$y = \frac{m_2}{n_2}.$$

Then take

$$z = \frac{m_1 n_2}{m_2 n_1},$$

whence

$$yz = \left(\frac{m_2}{n_2}\right)\left(\frac{m_2 n_2}{m_2 n_1}\right)$$

$$= \frac{m_2 m_1 n_2}{n_2 m_2 n_1}$$

$$= \left(\frac{m_1}{n_1}\right)\left(\frac{m_2 n_2}{m_2 n_2}\right)$$

$$\simeq \left(\frac{m_1}{n_1}\right)\left(\frac{1}{1}\right)$$

$$\simeq \frac{m_1 \cdot 1}{n_1 \cdot 1}$$

$$\simeq \frac{m_1}{n_1}$$

$$\simeq x. \qquad\qquad\qquad\qquad \text{Q.E.D.}$$

Some further obvious properties of fractions are stated in the exercise.

EXERCISES

1. Complete the proof of Theorem 1.
2. Prove Theorems 2, 3, and 4.
3. Prove Theorems 5, 6, and 7.
4. Prove Theorem 8.
5. Define $\leq_f$ in the obvious manner.
   (a) Is it antisymmetric?
   (b) Prove that if $x,y,u,v \in \mathrm{Fr}$ & $x \leq_f y$ & $x \simeq_f u$ & $y \simeq_f v$ then $u \leq_f v$.
   (c) Prove that if $x,y,z \in \mathrm{Fr}$ & $x \leq_f y$ & $y <_f z$ then $x <_f y$.
6. Give a direct conditional definition of addition of fractions without using a device like Definition 7.
7. Prove Theorems 9 and 10.
8. Prove Theorems 11 and 12.
9. Prove Theorems 13 and 14.
10. Prove Theorems 15 and 16.
11. Prove Theorem 17.

12. Do both addition and multiplication of fractions have the cancellation property (with respect to identity)?

13. Define $0_f = \frac{0}{1}$, and $1_f = \frac{1}{1}$. Prove that if $x,y,z,u,v \in \text{Fr}$, then:

   (a) $x + 0_f = x$,
   (b) $x \cdot 0_f = 0_f$,
   (c) $x \cdot 1_f = x$,
   (d)   if $xz <_f yz$ and $z >_f 0_f$ then $x <_f y$.
   (e)   if $y <_f z$ and $x >_f 0_f$ then $xy <_f xz$.

14. Give a counterexample to show that Theorem 18 is false if $\simeq_f$ is replaced by identity.

§ 6.3 Non-negative Rational Numbers.
We now develop the theory of non-negative rational numbers as equivalence classes of fractions.

DEFINITION 11. *If $x \in \text{Fr}$ then*

$$[x]_\nu = \{y: \ y \in \text{Fr} \ \& \ y \simeq_f x\}.$$

Thus $[x]_\nu$ is the set of all fractions equivalent to the fraction x. We use the subscript 'ν' to indicate 'non-negative rational'. (It is important to realize that the various subscripts introduced in this chapter like 'f' and 'ν' are not variables.)

DEFINITION 12. $\text{Nr} = \{A: \ (\exists x)(x \in \text{Fr} \ \& \ A = [x]_\nu\}.$

We have as an obvious result:

THEOREM 19. *The set Nr of non-negative rationals is a partition of the set Fr of fractions.*

We use capital letters 'M', 'N', 'P', 'Q', with or without subscripts, for non-negative rational numbers.

DEFINITION 13. $<_\nu = \{\langle M, N \rangle: \ M,N \in \text{Nr} \ \& \ (\exists x)(\exists y)(x \in M \ \& \ y \in N \ \& \ x <_f y)\}.$

We now use Theorem 7 to prove:

THEOREM 20. $<_\nu$ *is a strict simple ordering of* Nr.

PROOF. We need to show that $<_\nu$ is asymmetric, transitive, and connected in Nr.

[Asymmetry]. Assume $M,N \in \text{Nr}$ and $M <_\nu N$. Then there is a fraction x in M and a y in N such that

(1) $x <_f y$.

Suppose now there is a fraction u in M and a v in N such that

(2) $v <_f u$.

From Theorem 19 and familiar facts from Chapter 3 concerning partitions we know that

$$M = [x]_\nu = [u]_\nu,$$

$$N = [y]_\nu \quad [v]_\nu,$$

whence $x \simeq_f u$ and $y \simeq_f u$, and thus by virtue of Theorem 7 and (1)

$$u <_f v,$$

which is absurd, since $<_f$ is asymmetric (Theorem 3). Hence there are no fractions in M and N satisfying (2) and we conclude that it is not the case that $N <_\nu M$.

[Transitivity]. Assume: $M,N,P \in \mathrm{Nr}$ & $M <_\nu N$ & $N <_\nu P$. Then there is a fraction x in M, y_1 in N, y_2 in N, z in P, such that

$$x <_f y_1 \ \& \ y_2 <_f z,$$

but

$$y_1 \simeq_f y_2,$$

whence by Theorem 7

$$x <_f y_2,$$

and thus by virtue of the transitivity of $<_f$ (Theorem 3)

$$x <_f z,$$

from which we conclude:

$$M <_\nu P.$$

[Connectivity]. Assume $M,N \in \mathrm{Nr}$ and $M \neq N$. Since both M and N are non-empty, let x be an arbitrary element of M, and y of N. By virtue of Theorem 4 we have:

$$x \simeq_f y, \ x <_f y, \ \text{or} \ y <_f x,$$

but if $x \simeq_f y$ then $M = [x]_\nu = [y]_\nu = N$, contrary to our assumption, hence either $x <_f y$ or $y <_f x$, which establishes that $M <_\nu N$ or $N <_\nu M$. Q.E.D.

This proof illustrates the typical kind of argument used in going from one level of abstraction to a higher level. Everything hinges on two considerations: $<_f$ has most of the properties we expect $<_\nu$ to have, and $\simeq_f$ has substitution properties like identity.

We now define addition of non-negative rational numbers. To prove that all is as it should be we make crucial use of Theorem 9.

DEFINITION 14. $N + = \{\langle [x]_\nu, [y]_\nu, [z]_\nu\rangle : x,y,z \in \mathrm{Fr} \ \& \ x + y = z\}.$

THEOREM 21. *If* $M,N \in$ Nr *then there is a unique* P *in* Nr *such that* $\langle M,N,P \rangle \in N+$.

PROOF. By virtue of the definition of Nr there is an x in Fr and a y in Fr such that

$$M = [x]_\nu$$
$$N = [y]_\nu.$$

Since $x + y \in$ Fr, let

$$P = [x + y]_\nu.$$

Then $\langle M,N,P \rangle \in N+$.

To establish uniqueness of P, suppose $\langle M,N,P_1 \rangle \in N+$. Then by Definition 14 there must be elements u,v,w such that

$$M = [u]_\nu$$
$$N = [v]_\nu$$
$$P_1 = [w]_\nu$$
$$u + v = w.$$

But then

$$u \simeq_f x$$
$$v \simeq_f y,$$

whence by Theorem 9

$$u + v \simeq_f x + y,$$

that is,

$$w \simeq_f x + y,$$

from which we conclude:

$$P_1 = [w]_\nu = [x + y]_\nu = P,$$

as desired.   Q.E.D.

The theorem just proved justifies the definition of addition of non-negative rational numbers.

DEFINITION 15. *If* $M,N,P \in$ Nr *then* $M + N = P \leftrightarrow \langle M, N, P \rangle \in N+$.

THEOREM 22. *Addition of non-negative rational numbers is commutative and associative, has the cancellation property, and is monotonic with respect to the relation less than for non-negative rational numbers.*

proof. We only prove commutativity; the remaining parts of the proof follow a similar strategy.

There are fractions x and y such that

$$M = [x]_\nu$$
$$N = [y]_\nu,$$

whence

$$M + N = [x]_\nu + [y]_\nu$$
$$= [x + y]_\nu$$
$$= [y + x]_\nu$$
$$= [y]_\nu + [x]_\nu$$
$$= N + M. \qquad\qquad\text{Q.E.D.}$$

We now state without comment or proof the analogous definitions and theorems for multiplication of non-negative rational numbers.

Definition 16. $N\cdot = \{\langle [x]_\nu, [y]_\nu, [z]_\nu\rangle \colon x,y,z \in \text{Fr} \ \& \ xy = z\}.$

Theorem 23. *If $M,N \in \text{Nr}$ then there is a unique P in Nr such that $\langle M,N,P\rangle \in N\cdot$.*

Definition 17. *If $M, N, P \in \text{Nr}$ then $MN = P \leftrightarrow \langle M, N, P\rangle \in N\cdot$.*

Theorem 24. *Multiplication of non-negative rational numbers is commutative and associative, and it is distributive with respect to addition of non-negative rational numbers.*

In line with Exercise 13 of the preceding section we may define zero and one for non-negative rational numbers.

Definition 18. $0_\nu = [\tfrac{0}{1}]_\nu$

Definition 19. $1_\nu = [\tfrac{1}{1}]_\nu.$

Theorem 25. *We have:*

    (i) $0_\nu \neq 1_\nu,$
    (ii)   *if $M \in \text{Nr}$ then $M + 0_\nu = M$ and $M\cdot 1_\nu = M,$*
    (iii)   *if $M,N,P \in \text{Nr}$ and $0_\nu <_\nu M \ \& \ N <_\nu P$ then $MN <_\nu MP,$*
    (iv)   *if $M,N \in \text{Nr}$ and $0_\nu <_\nu N$ then there is a P in Nr such that $M = NP.$*

To introduce ordinary subtraction, we must construct the full set of rational numbers, positive, negative, and zero. This is the aim of the next section.

1. Prove Theorem 19.
2. Complete the proof of Theorem 22.
3. Prove that if $M,N \in$ Nr and $M <_\nu N$ then there is a unique P in Nr such that

$$M + P = N.$$

4. Prove Theorem 23.
5. Prove Theorem 24.
6. Prove Theorem 25.
7. Prove that if $M,N \in$ Nr and $0_\nu <_\nu M$ then there is an integer n such that

$$N <_\nu \left[\frac{n}{1} \right]_\nu \cdot M.$$

(This result establishes that the non-negative rational numbers have what is known as the Archimedean property.)

§ 6.4 Rational Numbers.

Before we can define the full set of rational numbers, we need entities which stand to the rational numbers in the way that fractions stand to the non-negative rational numbers. An ordered pair $\langle M, N \rangle$ of non-negative rational numbers is intuitively interpreted as $M - N$. The formal developments are very similar to those which have preceded, so the treatment will be rather summary. With subtraction in mind we label the set of pairs of non-negative rational numbers 'Sb'.

DEFINITION 20. $\mathrm{Sb} = \{\langle M, N \rangle : M,N \in \mathrm{Nr}\}$.

DEFINITION 21. $\simeq_\sigma = \{\langle\langle M_1,N_1 \rangle, \langle M_2,N_2 \rangle\rangle : M_1,N_1,M_2,N_2 \in \mathrm{Nr} \,\& \\ M_1 + N_2 = M_2 + N_1\}$.

THEOREM 26. $\simeq_\sigma$ *is an equivalence relation on* Sb.

DEFINITION 22. $<_\sigma = \{\langle\langle M_1,N_1 \rangle, \langle M_2,N_2 \rangle\rangle : M_1,N_1,M_2,N_2 \in \mathrm{Nr} \,\& \\ M_1 + N_2 <_\nu M_2 + N_1\}$.

We omit the obvious theorems on $<_\sigma$.

DEFINITION 23. $S+ = \{\langle\langle M_1,N_1 \rangle, \langle M_2,N_2 \rangle, \langle M_3,N_3 \rangle\rangle :$

$M_1,N_1,M_2,N_2,M_3,N_3 \in \mathrm{Nr} \,\& \langle M_1 + M_2, N_1 + N_2 \rangle \\ = \langle M_3,N_3 \rangle\}$.

THEOREM 27. *If* $M,N \in$ Sb *then there is a unique* P *in* Sb *such that* $\langle M, N, P \rangle \in S+$.

DEFINITION 24. *If* $M,N,P \in$ Sb *then* $M + N = P \leftrightarrow \langle M, N, P \rangle \in S+$.

DEFINITION 25. $S\cdot = \{\langle\langle M_1,N_1 \rangle, \langle M_2,N_2 \rangle, \langle M_3,N_3 \rangle\rangle : M_1,N_1,M_2,N_2,$

$M_3,N_3 \in \mathrm{Nr} \,\& \langle M_1 M_2 + N_1 N_2, M_1 N_2 + N_1 M_2 \rangle = \\ \langle M_3,N_3 \rangle\}$.

The simple intuitive idea of this rather complex defining condition is that

$$(M_1 - N_1)(M_2 - N_2) = M_3 - N_3 \leftrightarrow M_1M_2 - M_1N_2 - N_1M_2 + N_1N_2$$
$$= M_3 - N_3.$$

Theorem 28. *If* $M,N \in$ Sb *then there is a unique* P *in* Sb *such that* $\langle M,N,P \rangle \in S\cdot$.

Definition 26. *If* $M,N,P \in$ Sb *then* $MN = P \leftrightarrow \langle M,N,P \rangle \in S\cdot$.

Theorem 29. *If* $M_1, M_2, N_1, N_2 \in$ Sb $\&$ $M_1 \simeq_\sigma M_2$ $\&$ $N_1 \simeq_\sigma N_2$, *then*:

(i)   *If* $M_1 <_\sigma N_1$ *then* $M_2 <_\sigma N_2$,
(ii) $M_1 + N_1 \simeq_\sigma M_2 + N_2$,
(iii) $M_1 N_1 \simeq_\sigma M_2 N_2$.

In brief, this theorem asserts that the equivalence relation defined for Sb has the appropriate substitution properties. Rather than state properties of the operations on ordered pairs of non-negative rational numbers, we go on to define equivalence classes of these pairs and thus obtain the rational numbers. Following Whitehead and Russell, we use the subscript '$_s$' for rational number operations and relations, reserving '$_r$' for the real numbers. Also, for purposes of later work we informally use general variables 'x', 'y', 'z', 'u', 'v', with or without subscripts, for rational numbers. Explicit reference to the set Ra of rationals will avoid any ambiguities.

Definition 27. *If* $M \in$ Sb *then* $[M]_s = \{N \colon N \in$ Sb $\& N \simeq_\sigma M\}$.

Definition 28. Ra $= \{x \colon (\exists M)(M \in$ Sb $\& x = [M]_s)\}$.

Before stating one comprehensive theorem on rational numbers we need to define *less than, addition,* and *multiplication.*

Definition 29. $<_s = \{\langle [M]_s, [N]_s \rangle \colon M,N \in$ Sb $\& M <_\sigma N\}$.

Definition 30. $R+ = \{\langle [M]_s, [N]_s, [P]_s \rangle \colon M,N,P \in$ Sb $\& M + N = P\}$.

Theorem 30. *If* x *and* y *are rational numbers* (*i.e.,* $x,y \in$ Ra) *then there is a unique rational number* z *such that* $\langle x, y, z \rangle \in R+$.

Definition 31. *If* x,y, *and* z *are rational numbers then*
$$x + y = z \leftrightarrow \langle x, y, z \rangle \in R+.$$

Definition 32. $R\cdot = \{\langle [M]_s, [N]_s, [P]_s \rangle \colon M,N,P \in$ Sb $\& MN = P\}$.

Theorem 31. *If* x *and* y *are rational numbers then there is a unique rational number* z *such that* $\langle x, y, z \rangle \in R\cdot$.

Definition 33. *If* x,y *and* z *are rational numbers then*
$$xy = z \leftrightarrow \langle x, y, z \rangle \in R\cdot.$$

We also define zero and one.

DEFINITION 34.

$$0_s = [\langle 0_\nu, 0_\nu \rangle]_s$$

$$1_s = [\langle 1_\nu, 0_\nu \rangle]_s$$

THEOREM 32. *The relation less than and the operations of addition and multiplication for rational numbers together with 0_s and 1_s have the following fifteen properties for all rational numbers x, y, and z:*

(1) $x + y = y + x$,
(2) $xy = yx$,
(3) $(x + y) + z = x + (y + z)$,
(4) $(xy)z = x(yz)$,
(5) $x(y + z) = xy + xz$,
(6) $x + 0 = x$,
(7) $x \cdot 1 = x$,
(8) *There is a rational number y such that $x + y = 0$,*
(9) *If $y \neq 0$ then there is a rational number z such that $x = yz$,*
(10) *If $x <_s y$ then not $y <_s x$,*
(11) *If $x <_s y$ and $y <_s z$ then $x <_s z$,*
(12) *If $x \neq y$ then $x <_s y$ or $y <_s x$,*
(13) *If $y <_s z$ then $x + y <_s x + z$,*
(14) *If $0 <_s x$ and $y <_s z$ then $xy <_s xz$,*
(15) $0_s \neq 1_s$.

The proof of the fifteen parts of the theorem we leave as an exercise. From this point on we assume all familiar arithmetical results about rational numbers without further explicit development.*

Assuming the negative of a rational number to be defined, and $\leq_s$, it is also desirable because of its importance for the next section to define the absolute value of a rational number and summarize the properties of this operation.

DEFINITION 35. *If x is a rational number then*

$$|x| = y \leftrightarrow (x \geq_s 0 \rightarrow y = x) \ \& \ (x <_s 0 \rightarrow y = -x).$$

In the usual mathematical notation we would write:

$$|x| = \begin{cases} x \text{ if } x \geq 0, \\ -x \text{ if } x < 0. \end{cases}$$

---

*Deductive developments from the fifteen properties of Theorem 32 are to be found in my *Introduction to Logic*.

Theorem 33. *If x and y are rational numbers then:*

    (i) $|x| \geq 0$
    (ii) $|xy| = |x| \cdot |y|$
    (iii) $|x + y| \leq |x| + |y|$
    (iv) $|x| - |y| \leq |x - y|$
    (v) $x \cdot |y| \leq |xy|$.

To indicate how many levels of abstraction we have gone through, we define those rational numbers corresponding to the natural numbers.

Definition 36. $n_s = \left[\left\langle \left[\dfrac{n}{1} \right]_\nu , \left[\dfrac{0}{1} \right]_\nu \right\rangle \right]_s$.

The facts expressed in the final theorem of this section are both intrinsically interesting and useful later. The proof uses the diagonal method of Cauchy for rearranging a doubly infinite sequence into a simply infinite one (compare the proof in §5.3 that $\omega \times \omega \approx \omega$).

Theorem 34. *The set of rational numbers is denumerable and can be well-ordered without using the axiom of choice.*

proof. It will be obvious that the ordering constructed establishes both parts of the proof. Details are omitted. We represent each rational number by a fraction m/n such that m and n have no common factor greater than 1. Thus the rational number one-half is represented by $\frac{1}{2}$, not $\frac{2}{4}$ or $\frac{3}{6}$. Negative one-half by $-\frac{1}{2}$, zero by $\frac{0}{1}$, etc. (Translation of these intuitive notions into the equivalence class constructions actually used to obtain the rational numbers is routine.) We then arrange the fractions, first according to the sum of m and n. If $m + n = m' + n'$, m/n precedes m'/n' when $m < m'$. Finally, negative fractions immediately precede their positive counterparts. The ordering then is:

$$\frac{0}{1}, -\frac{1}{1}, \frac{1}{1}, -\frac{1}{2}, \frac{1}{2}, -\frac{2}{1}, \frac{2}{1}, -\frac{1}{3}, \frac{1}{3}, -\frac{3}{1}, \frac{3}{1}, -\frac{1}{4}, \frac{1}{4}, -\frac{2}{3}, \frac{2}{3}, \ldots \quad \text{Q.E.D.}$$

To avoid continual use of subscripts in subsequent sections we shall not distinguish between n_s and n, and the context should make plain which entity is appropriate. One useful definition may be framed, which refers both to n and n_s. This is the definition of the least integer equal to or greater than a positive rational number.

Definition 37. *If x is a positive rational number then*

$$[x] = n \leftrightarrow x \leq n_s \ \& \ (\forall m)(x \leq m_s \rightarrow n_s \leq m_s).$$

Thus in usual intuitive notation:*

$$\left[\frac{1}{2}\right] = 1$$

$$[2] = 2$$

$$\left[1\frac{1}{7}\right] = 2.$$

EXERCISES

1. Prove Theorem 26.
2. Prove Theorems 27 and 28.
3. Prove Theorem 29.
4. Prove Theorems 30 and 31.
5. Prove Theorem 32. (This exercise has fifteen parts.)
6. Prove Theorem 33.
7. Give a detailed proof of Theorem 34.
8. Prove that if x and y are positive rational numbers then:
    (i) $[x + y] \leq [x] + [y]$,
    (ii) $[xy] \leq [x] [y]$.

§ 6.5 Cauchy Sequences of Rational Numbers.

We now develop the fundamental facts about Cauchy sequences of rational numbers on the basis of which we construct the real numbers in the next section. The subscript notation '$_s$' is dropped in this section and the remainder of the chapter.

DEFINITION 38. *x is a sequence if and only if x is a function on the set ω of natural numbers.*

We introduce the usual subscript notation for *members* of a sequence.

DEFINITION 39.† *If x is a sequence, $x_n = x(n)$.*

In the usual terminology x_n is the *n-th member* or *term* of the sequence x.

Slightly in violation of our standard rules for definition, we also introduce a customary notation for sequences.

DEFINITION 40. *If x is a sequence, $\langle x_1, x_2, \ldots, x_n, \ldots \rangle = x$,*

and the sequences we are particularly interested in are:

---

*This additional use of square braces should not be confusing, since no subscript is used here. It is actually more common to define $[x]$ as the largest integer $\leq x$. The present notation is very convenient for our purposes.

†This definition essentially duplicates Definition 15 of §5.2.

DEFINITION 41.   *x is a sequence of rational numbers if and only if x is a sequence and the range of x is a subset of the set of rational numbers.*

Operations on sequences of rational numbers are defined in the expected way. The line of reasoning justifying the definitions is too obvious to require separate theorems.

DEFINITION 42.   *If x and y are sequences of rational numbers then*

$$x + y = z \leftrightarrow (\forall n)(x_n + y_n = z_n).$$

DEFINITION 43.   *If x and y are sequences of rational numbers then*

$$xy = z \leftrightarrow (\forall n)(x_n y_n = z_n).$$

Clearly if x and y are sequences of rational numbers, then so are $x + y$ and xy.

Proofs of the expected properties of these operations of addition and multiplication follow easily from the properties of the corresponding operations for rational numbers.

THEOREM 35.   *Addition of sequences of rational numbers is commutative and associative and has the cancellation property.*

PROOF.   We only prove right-hand cancellation. Let x,y,z be sequences of rational numbers.

$$x + z = y + z \leftrightarrow (\forall n)(x_n + z_n = y_n + z_n)$$

$$\leftrightarrow (\forall n)(x_n = y_n)$$

$$\leftrightarrow x = y. \qquad \text{Q.E.D.}$$

THEOREM 36.   *Multiplication of sequences of rational numbers is commutative, associative, and distributive with respect to addition.*

The sequences of rational numbers essential to construction of the real numbers are the Cauchy sequences now to be defined.

DEFINITION 44.   *x is a Cauchy sequence of rational numbers if and only if x is a sequence of rational numbers, and for every rational number $\epsilon > 0$ there is a positive integer N such that for every $m,n > N$*

$$|x_n - x_m| < \epsilon.*$$

As remarked in the introduction to this chapter Cauchy sequences are also called *convergent sequences, regular sequences,* and *fundamental sequences.*

---

*We use 'ϵ' in deference to traditional notation, although it violates our previous convention that lower case Greek letters are variables taking ordinal numbers as values. For a similar reason we often use 'positive integer' or simply 'integer' in place of 'natural number'.

Some examples of Cauchy sequences will not be amiss in view of the fundamental importance of the notion. Let x be the sequence of rational numbers such that $x_0 = 0$ and for $n \geq 1$

$$x_n = \frac{1}{n}.$$

To show that x is a Cauchy sequence we need to find for each positive rational number ϵ, an integer N such that for $m, n > N$

$$|x_n - x_m| < \epsilon.$$

Here the appropriate N is easy to find. Let

$$N = \left[\frac{1}{\epsilon} \right].$$

Then the following inequalities for $m, n > [\frac{1}{\epsilon}]$ establish that x is a Cauchy sequence:

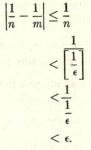

$$\left| \frac{1}{n} - \frac{1}{m} \right| \leq \frac{1}{n}$$
$$< \frac{1}{\left[\frac{1}{\epsilon} \right]}$$
$$< \frac{1}{\frac{1}{\epsilon}}$$
$$< \epsilon.$$

As a second example, let x be the sequence such that for $n > 0$

$$x_n = \frac{n^2 - 1}{n^2}.$$

Here the natural choice of N is $\left[\dfrac{1}{\sqrt{\epsilon}} \right]$, but $\sqrt{\epsilon}$ is not in general a rational number. An appropriate rational number choice we leave as an exercise.

We now want to prove the important fact that the sum and product of two Cauchy sequences is also a Cauchy sequence. The proof concerning the product of two such sequences makes use of the following fact.

THEOREM 37. *If x is a Cauchy sequence of rational numbers then there is a positive rational number δ such that for every n*

$$|x_n| < \delta.$$

PROOF. By virtue of the fact that x is a Cauchy sequence there is an integer N such that for every $m, n > N$

$$(1) \qquad\qquad |x_n - x_m| < 1.$$

Let

(2) $$\delta = \max (|x_0|, |x_1|, \ldots, |x_N|, |x_{N+1}|) + 1.$$

Obviously for $n \leq N + 1$

$$|x_n| < \delta.$$

Suppose then $n > N + 1$. By virtue of **(1)**

$$|x_n| < |x_{N+1}| + 1,$$

but in view of (2)

$$|x_{N+1}| + 1 \leq \delta. \qquad \text{Q.E.D.}$$

THEOREM 38. *If x and y are Cauchy sequences of rational numbers, then $x + y$ and xy are also Cauchy sequences of rational numbers.*

PROOF. [Sum]. Let $\epsilon > 0$. By hypothesis there are numbers M and N such that if $m,n > M$ then

$$\left| x_n - x_m \right| < \frac{\epsilon}{2},$$

and if $m,n > N$

$$\left| y_n - y_m \right| < \frac{\epsilon}{2}$$

Let

$$P = \max (M,N).$$

If $m,n > P$ then

$$|(x_n + y_n) - (x_m + y_m)| = |(x_n - x_m) + (y_n - y_m)|$$
$$\leq |x_n - x_m| + |y_n - y_m|$$
$$< \frac{\epsilon}{2} + \frac{\epsilon}{2}$$
$$< \epsilon.$$

[Product]. Let $\epsilon > 0$. By virtue of the preceding theorem there are positive rational numbers δ_1 and δ_2 such that for every n

$$|x_n| < \delta_1$$
$$|y_n| < \delta_2.$$

Let $\delta = \max (\delta_1, \delta_2)$. Moreover, since x and y are Cauchy sequences there is an integer M_1 such that for $m,n > M_1$

$$|x_n - x_m| < \epsilon /(2\delta),$$

and there is an integer M_2 such that for $m,n > M_2$

$$|y_n - y_m| < \epsilon /(2\delta).$$

Let $M = \max (M_1, M_2)$. Then for $m,n > M$

$$|x_n y_n - x_m y_m| = |x_n y_n - x_n y_m + x_n y_m - x_m y_m|$$

$$\leq |x_n|\,|y_n - y_m| + |y_m|\,|x_n - x_m|$$

$$< \delta \left(\frac{\epsilon}{2\delta}\right) + \delta \left(\frac{\epsilon}{2\delta}\right)$$

$$< \epsilon. \qquad\qquad \text{Q.E.D.}$$

In our development real numbers are certain equivalence classes of Cauchy sequences of real numbers; we now define the appropriate equivalence relation.

DEFINITION 45. *If x and y are Cauchy sequences of rational numbers then*: $x \simeq_c y$ *if and only if for every positive rational number ϵ there is an integer N such that for every $n > N$*

$$|x_n - y_n| < \epsilon.$$

Note that two Cauchy sequences can be equivalent even if they differ in every term. For example, if for $n > 0$

$$x_n = \frac{n^2 - 1}{n^2},$$

$$y_n = \frac{n^2 + 1}{n^2}.$$

Then for every $n > 0$, $x_n \neq y_n$, but $x \simeq_c y$.

Since we have not defined the relation $\simeq_c$ as a set-theoretical entity, the appropriate theorem concerning its properties as an equivalence relation must be spelled out. The proof is left as an exercise.

THEOREM 39. *If x, y and z are Cauchy sequences of rational numbers then*:

(i) $x \simeq_c x$,

(ii)   *if $x \simeq_c y$ then $y \simeq_c x$,*

(iii) *if $x \simeq_c y$ and $y \simeq_c z$ then $x \simeq_c z$.*

Analogous to equivalence is the relation *less than* for Cauchy sequences.

DEFINITION 46. *If x and y are Cauchy sequences of rational numbers then*: $x <_c y$ *if and only if there is a rational number $\delta > 0$ and an integer N such that for $n > N$*

$$y_n > x_n + \delta.$$

The expected properties hold.

THEOREM 40. *If x and y are Cauchy sequences of rational numbers then exactly one of the following holds: $x \simeq_c y$, $x <_c y$, $y <_c x$.*

PROOF. It is obvious from the definitions given that at most one of the relationships can hold. So we need to prove that at least one holds.

Suppose that x is not equivalent to y. It follows immediately from Definition 45 that there is a positive rational number 2ϵ such that for every integer N there is an $n > N$ such that

$$(1) \qquad\qquad |x_n - y_n| > 2\,\epsilon.$$

But since x and y are Cauchy sequences there is an integer M_1 such that if $m,n > M_1$ then

$$\left| x_n - x_m \right| < \frac{\epsilon}{2},$$

and there is an integer M_2 such that if $m,n > M_2$, then

$$\left| y_n - y_m \right| < \frac{\epsilon}{2}$$

Using the result which led to (1), there is an integer p such that

$$p > \max\,(M_1, M_2)$$

and

$$(2) \qquad\qquad |x_p - y_p| > 2\,\epsilon.$$

We now have two possibilities: $x_p > y_p$ or $y_p > x_p$.
Suppose $x_p > y_p$. Then by virtue of (2)

$$(3) \qquad\qquad x_p > y_p + 2\,\epsilon.$$

Moreover, for every $n > p$

$$(4) \qquad\qquad \left| x_p - x_n \right| < \frac{\epsilon}{2}$$

$$(5) \qquad\qquad \left| y_p - y_n \right| < \frac{\epsilon}{2},$$

whence from (3) and (4)

$$(6) \qquad x_n > x_p - \frac{\epsilon}{2} > y_p + 2\epsilon - \frac{\epsilon}{2} = y_p + \frac{3\epsilon}{2}.$$

But it follows from (5) that

$$(7) \qquad\qquad y_p > y_n - \frac{\epsilon}{2},$$

and we infer from (6) and (7) that for every $n > p$

$$x_n > y_n - \frac{\epsilon}{2} + \frac{3\epsilon}{2} = y_n + \epsilon.$$

Thus on our supposition $y <_c x$.

If $y_p > x_p$ we infer by the same kind of reasoning that $x <_c y$. Q.E.D.
The proof that *less than* for Cauchy sequences is asymmetric and transitive is left as an exercise.

Theorem 41. *If x, y, and z are Cauchy sequences of rational numbers then:*

(i)   *if $x <_c y$ then not $y <_c x$,*
(ii) *if $x <_c y$ and $y <_c z$ then $x <_c z$.*

The final theorem of this section states that the equivalence relation defined for Cauchy sequences has the appropriate substitution properties.

Theorem 42. *If x, y, u, and v are Cauchy sequences of rational numbers and $x \simeq_c u$ and $y \simeq_c v$, then:*

(i)   *if $x <_c y$ then $u <_c v$,*
(ii) $x + y \simeq_c u + v$,
(iii) $xy \simeq_c uv$.

proof. We prove only (iii). Let $\epsilon > 0$. By Theorem 37 there are positive rational numbers δ_1, δ_2, δ_3, δ_4 such that for every n

$$|x_n| < \delta_1$$
$$|y_n| < \delta_2$$
$$|u_n| < \delta_3$$
$$|v_n| < \delta_4.$$

Let $\delta = \max(\delta_1, \delta_2, \delta_3, \delta_4)$. By hypothesis of equivalence there is an integer N_1 such that for $n > N_1$

$$|x_n - u_n| < \epsilon/2\delta,$$

and there is an integer N_2 such that for $n > N_2$

$$|y_n - v_n| < \epsilon/2\delta.$$

Let $N = \max(N_1, N_2)$. Then for every $n > N$, we have:

$$|x_n y_n - u_n v_n| = |x_n y_n - u_n y_n + u_n y_n - u_n v_n|$$
$$\leq |y_n|\,|x_n - u_n| + |u_n|\,|y_n - v_n|$$
$$< \delta(\epsilon/2\delta) + \delta(\epsilon/2\delta)$$
$$< \epsilon. \qquad\qquad \text{Q.E.D.}$$

Note how similar this proof is to the proof that the product of two Cauchy sequences is a Cauchy sequence.

EXERCISES

1. Prove that the set of all sequences of rational numbers exists.

2. Let x be the sequence such that

$$x_0 = 0$$

$$x_n = \frac{n^2 - 1}{n^2} \text{ for } n > 0.$$

Prove that x is a Cauchy sequence.

3. Let x be the sequence such that

$$x_n = \frac{2^n}{n!}.$$

Prove that x is a Cauchy sequence.

4. Complete the proof of Theorem 35.

5. Prove Theorem 36.

6. Prove Theorem 39.

7. Prove Theorem 41.

8. Complete the proof of Theorem 42.

9. First we define: x is a *monotone increasing sequence of integers* if and only if (i) x is a sequence, (ii) the range of x is a subset of ω, (iii) if $m < n$ then $x_m < x_n$. Next we define: x is a *subsequence* of y if and only if y is a sequence and there exists a monotone increasing sequence of integers z such that $x = y \circ z$. (This provides a precise definition of the intuitively familiar idea of a subsequence. The symbol $\circ$ designates composition of functions.)

   (a)   Give an example of a sequence of rational numbers which is not a Cauchy sequence, but which has a subsequence which is.

   (b)   Give an example of a sequence of rational numbers which has no subsequence which is a Cauchy sequence.

   (c)   Prove that every subsequence of a Cauchy sequence of rational numbers is also a Cauchy sequence of rational numbers.

   (d)   Prove that if y is a Cauchy sequence of rational numbers and x is a subsequence of y then $x \simeq_c y$.

§ 6.6 Real Numbers.

We begin by defining equivalence classes of Cauchy sequences of rational numbers.

DEFINITION 47.   *If x is a Cauchy sequence of rational numbers then* $[x]_r = \{y: y$ *is a Cauchy sequence of rational numbers* $\& y \simeq_c x\}$.

And we then define the set Re of real numbers.

DEFINITION 48.   Re $= \{y: (\exists x)(x$ *is a Cauchy sequence of rational numbers* $\& y = [x]_r)\}$.

We have the obvious theorem:

THEOREM 43.   *The set of real numbers is a partition of the set of Cauchy sequences of rational numbers.*

We next define less than, addition, and multiplication.

DEFINITION 49. $<_r = \{\langle [x]_r, [y]_r \rangle: x, y$ are Cauchy sequences of rational numbers & $x <_c y\}$.

DEFINITION 50. $R+ = \{\langle [x]_r, [y]_r, [z]_r \rangle: x,y,z$ are Cauchy sequences of rational numbers & $x + y = z\}$.

THEOREM 44. If x and y are real numbers then there is a unique real number z such that $\langle x, y, z \rangle \in R+$.

DEFINITION 51. If $x,y,z \in$ Re then $x + y = z \leftrightarrow \langle x, y, z \rangle \in R+$.

DEFINITION 52. $R\cdot = \{\langle [x]_r, [y]_r, [z]_r \rangle: x,y,z$ are Cauchy sequences of rational numbers & $xy = z\}$.

THEOREM 45. If x and y are real numbers then there is a unique real number z such that $\langle x, y, z \rangle \in R\cdot$.

DEFINITION 53. If $x,y,z \in$ Re then $xy = z \leftrightarrow \langle x, y, z \rangle \in R\cdot$.

We also define the real numbers zero and one.

DEFINITION 54. $0_r = [\langle 0_s, 0_s, \ldots, 0_s, \ldots \rangle]_r$.

DEFINITION 55. $1_r = [\langle 1_s, 1_s, \ldots, 1_s, \ldots \rangle]_r$.

Our comprehensive theorem on elementary properties of the real numbers is like Theorem 32 for the rational numbers. Since all parts of the proof closely resemble proofs of previous theorems, nothing is proved here.

THEOREM 46. *The relation less then and the operations of addition and multiplication for real numbers together with 0_r and 1_r have all the fifteen properties listed in Theorem 32.*

To complete the construction of the real numbers we need only prove that they satisfy one further property: every non-empty set of real numbers which has an upper bound has a least upper bound.* However, the proof that this property is satisfied requires some other results which are themselves important and of intrinsic interest. We begin with the notion of a real number *corresponding to* a rational number.

DEFINITION 56. *If x is a real number and s is a rational number then x corresponds to s if and only if the sequence $\langle s, s, \ldots, s, \ldots \rangle$ is a member of x.*

We have as an immediate consequence of the fact that the set of real numbers is a partition of the set of Cauchy sequences of rational numbers the result that:

---

*That this property and the fifteen of Theorem 32 adequately characterize the real numbers is well-known; a discussion of these matters is to be found in nearly any book on the theory of functions of a real variable.

THEOREM 47. *Given any rational number there is a unique real number corresponding to it.*

Combining this theorem with Theorem 34 we obtain:

THEOREM 48. *The set of real numbers corresponding to rational numbers is denumerable and can be well-ordered without using the axiom of choice.*

A useful property of the real numbers corresponding to rational numbers is formulated in the next theorem: briefly, the rational real numbers, as we may appropriately call them, are dense in the set of all real numbers.*

THEOREM 49. *If x and y are any two distinct real numbers then there exists a real number z corresponding to some rational number such that z is between x and y, i.e., if $x < y$ then $x < z$ and $z < y$, and if $y < x$ then $y < z$ and $z < x$.*

PROOF. For definiteness, let $x < y$. Let $a \in x$ and $b \in y$, whence a and b are Cauchy sequences of rational numbers with

$$(1) \qquad\qquad a < b.$$

From (1), Definition 46, and Definition 42 we infer that there is a positive rational number δ and an integer N such that for all $m,n > N$

$$(2) \qquad\qquad b_n - a_n > \delta$$

$$(3) \qquad\qquad |a_n - a_m| < \delta/4$$

$$(4) \qquad\qquad |b_n - b_m| < \delta/4.$$

Let s be a rational number such that

$$(5) \qquad\qquad \delta/4 < s < \delta/2,$$

and consider the rational number $a_{N+1} + s$. I say that the real number z corresponding to $a_{N+1} + s$ is between x and y. (Here $z = [\langle a_{N+1} + s, a_{N+1} + s, \ldots, a_{N+1} + s, \ldots \rangle]_r$.)

First, we infer from (3) that

$$a_n - a_{N+1} < \delta/4$$

whence multiplying through by -1 and adding s to both sides we obtain:

$$(a_{N+1} + s) - a_n > s - \delta/4,$$

from which we conclude by Definition 46 that

$$\langle a_{N+1} + s, a_{N+1} + s, \ldots, a_{N+1} + s, \ldots \rangle > a,$$

and thence from Definition 49

$$z > x.$$

*In the remainder of this chapter we omit the subscript 'r' in '$<_r$' and elsewhere.

Similarly, from (2) and (4) we infer

$$b_m - (a_{N+1} + s) = (b_{N+1} - a_{N+1}) + (b_m - b_{N+1}) - s > \delta - \delta/4 - s.$$

By virtue of (5) we know that $\delta - \delta/4 - s$ is positive and we conclude that $y >_r z$. Q.E.D.

We define Cauchy sequences of real numbers just as we define such sequences of rational numbers.

> **DEFINITION 57.** *x is a sequence of real numbers if and only if x is a sequence and the range of x is a subset of the set of real numbers.*

> **DEFINITION 58.** *x is a Cauchy sequence of real numbers if and only if x is a sequence of real numbers and for every real number $\epsilon > 0$ there is a positive integer N such that for every m,n > N*
>
> $$|x_n - x_m| < \epsilon.$$

The proof of the first theorem about Cauchy sequences of real numbers is not difficult and we leave it as an exercise.

> **THEOREM 50.** *If a is a Cauchy sequence of rational numbers and x is the sequence of real numbers such that for every n, x_n is the real number corresponding to a_n, then x is a Cauchy sequence of real numbers. Conversely, if x is a Cauchy sequence then so is a.*

We now define the important notion of limit. Without question this notion is *the* fundamental concept of the differential and integral calculus, and of analysis in general. What we want to show is that a sequence of real numbers has a limit if and only if it is a Cauchy sequence. This result is sometimes called the General Convergence Principle. In more recent terminology it would be called the theorem on the completeness of the real number system. Independent of terminology, the important fact is that this result, like the least upper bound property, expresses the essential difference between the real numbers and the rational numbers. It is easy to construct examples of Cauchy sequences of rational numbers which do not have a limit. For example, let $x_0 = 0$ and for $n \geq 1$

$$x_n = \left(1 + \frac{1}{n}\right)^n.$$

It is easy to see that x is a Cauchy sequence of rational numbers, yet it does not have a limit among the rational numbers. (Its limit is, in fact, the base e for natural logarithms: $e = 2.71828\ldots$.)

> **DEFINITION 59.** *If x is a sequence of real numbers, then y is a limit of x if and only if y is a real number and for every real number $\epsilon > 0$ there is a positive integer N such that for every n > N*
>
> $$|x_n - y| < \epsilon.$$

The proof is straightforward that

THEOREM 51.   *A sequence of real numbers has at most one limit.*

This theorem justifies definition of the standard notation for limit of a sequence.

DEFINITION 60.   *If x is a sequence of real numbers and y is the limit of x then*

$$\lim_{n \to \infty} x_n = y.$$

THEOREM 52.   *A sequence of real numbers has a limit if and only if it is a Cauchy sequence.*

PROOF. [Necessity]. Let x be a sequence of real numbers whose limit is y. Let ϵ be any positive real number. Then by Definition 59 there is an integer N such that for $m,n > N$

$$|y - x_n| < \epsilon/2$$
$$|y - x_m| < \epsilon/2,$$

whence

$$|x_n - x_m| \leq |y - x_n| + |y - x_m| < \epsilon/2 + \epsilon/2 = \epsilon,$$

and we conclude from Definition 58 that x is a Cauchy sequence of real numbers.

[Sufficiency]. Let x be a Cauchy sequence of real numbers. Let W be the well-ordering of the rational real numbers corresponding to that described for the rational numbers in the proof of Theorem 34.

Let y_n be the W-first rational real number between x_n and $x_n + \dfrac{1}{n}$, i.e.,

$$x_n < y_n < x_n + \frac{1}{n},$$

whence for every positive real number ϵ there is an integer N such that for every $n > N$

$$(1) \qquad \left| x_n - y_n \right| < \frac{1}{N} < \epsilon/3.$$

Moreover, y is a Cauchy sequence of real numbers, for

$$|y_n - y_m| < |y_n - x_n| + |x_n - x_m| + |x_m - y_m|,$$

and N may be so chosen that for $m,n > N$

$$|y_n - x_n| < \epsilon/3$$
$$|x_m - y_m| < \epsilon/3$$
$$|x_n - x_m| < \epsilon/3,$$

hence

$$|y_n - y_m| < \epsilon.$$

Now let a_n be the rational number to which y_n corresponds. By Theorem 50 a is a Cauchy sequence of rational numbers, and thus $[a]_r$ is a real number. From Definition 45 it then easily follows that for every positive real number ϵ there is an integer N such that for $n > N$

(2) $$|y_n - [a]_r| < \epsilon/2,$$

whence, combining (1) and (2), N may be chosen so that for $n > N$

$$|x_n - [a]_r| \leq |x_n - y_n| + |y_n - [a]_r| < \epsilon/2 + \epsilon/3 < \epsilon,$$

which establishes that $[a]_r$ is the limit of x. Q.E.D.

It should be remarked that the introduction of the well-ordering W of the rational numbers in the second half of the above proof is unorthodox. The standard proofs to be found in most textbooks on functions of real variables read: *Select a y_n such that*

$$x_n < y_n < x_n + \frac{1}{n},$$

but an infinite number of such selections is required and thus the standard proofs depend on the axiom of choice. A great many proofs in analysis which depend on the axiom of choice may be modified in the way this proof has been modified to avoid such dependence.

The fact that every Cauchy sequence of real numbers has a limit shows that we cannot essentially extend the number system further by defining what we may call super-real numbers as equivalence classes of Cauchy sequences of real numbers. Super-real numbers have just the properties of real numbers. To each real number there corresponds a unique super-real number, and conversely. Moreover, this uniqueness relation is preserved under the operations of addition and multiplication, and under limiting operations as well.

We now proceed as directly as possible to the least upper bound theorem.

Definition 61. *If A is a set of real numbers then y is an upper bound of A if and only if y is a real number and for every x in A, $x \leq y$.*

Definition 62. *If A is a set of real numbers then y is a least upper bound of A if and only if y is an upper bound of A and for every z which is an upper bound of A, $y \leq z$.*

We leave proofs of the next two theorems as exercises, as well as the proof that an empty set of real numbers has no least upper bound.

Theorem 53. *If A is a set of real numbers then y is a least upper bound of A if and only if y is an upper bound of A and for every positive real number ϵ there is an x in A such that $x > y - \epsilon$.*

Theorem 54. *A set of real numbers can have at most one least upper bound.*

We are now prepared to prove:

Theorem 55. *If a non-empty set of real numbers has an upper bound then it has a least upper bound.*

proof. Let A be a set satisfying the hypothesis of the theorem. Clearly there are two real numbers x_0 and u_0 such that $u_0 - x_0 = 1$, u_0 is an upper bound of A, and x_0 is not. We now define inductively x_n and u_n

$$x_1 = \begin{cases} \dfrac{x_0 + u_0}{2} & \text{if } \dfrac{x_0 + u_0}{2} \text{ is not an upper bound of } A, \\[2mm] x_0 & \text{if } \dfrac{x_0 + u_0}{2} \text{ is an upper bound of } A. \end{cases}$$

$$u_1 = \begin{cases} \dfrac{x_0 + u_0}{2} & \text{if } \dfrac{x_0 + u_0}{2} \text{ is an upper bound of } A, \\[2mm] u_0 & \text{if } \dfrac{x_0 + u_0}{2} \text{ is not an upper bound of } A. \end{cases}$$

And, in general

$$x_n = \begin{cases} \dfrac{x_{n-1} + u_{n-1}}{2} & \text{if } \dfrac{x_{n-1} + u_{n-1}}{2} \text{ is not an upper bound of } A, \\[2mm] x_{n-1} & \text{if } \dfrac{x_{n-1} + u_{n-1}}{2} \text{ is an upper bound of } A. \end{cases}$$

$$u_n = \begin{cases} \dfrac{x_{n-1} + u_{n-1}}{2} & \text{if } \dfrac{x_{n-1} + u_{n-1}}{2} \text{ is an upper bound of } A, \\[2mm] u_{n-1} & \text{if } \dfrac{x_{n-1} + u_{n-1}}{2} \text{ is not an upper bound of } A. \end{cases}$$

We then have:

(1) $$u_n - x_n = \frac{1}{2^n}$$
and
$$u_{n-1} - u_n \le \frac{1}{2^n},$$

from which we easily infer that the sequence u is a Cauchy sequence of real numbers, for given $\epsilon > 0$ there is an N such that for every $m,n > N$ with $n > m$

$$\frac{1}{2^{m+1}} + \cdots + \frac{1}{2^n} < \epsilon,$$

whence

$$\left| u_m - u_n \right| = \left| u_m - u_{m+1} \right| + \left| u_{m+1} - u_{m+2} \right| + \cdots + \left| u_{n-1} - u_n \right|$$
$$\leq \frac{1}{2^{m+1}} + \cdots + \frac{1}{2^n} < \epsilon.$$

And by Theorem 52 the sequence u has a limit, say $y\star$. We want to show that $y\star$ is the least upper bound of A. First, we establish that $y\star$ is an upper bound of A. Let $x \in A$. For every n, $u_n \geq x$. Since for every ϵ there is an n such that

$$u_n - y\star < \epsilon,$$

we have

$$y\star + \epsilon > u_n,$$

whence

(2) $$y\star + \epsilon > x,$$

but since (2) must hold for all $\epsilon > 0$,

$$y\star \geq x.$$

Thus $y\star$ is an upper bound of A.

Now suppose z is an upper bound of A such that $z < y\star$. Then on the basis of (1) there is an n such that

$$u_n - x_n < y\star - z,$$

but

$$y\star - x_n \leq u_n - x_n,$$

whence

$$y\star - x_n < y\star - z,$$

and thus

$$z < x_n,$$

but then z is not an upper bound of A since x_n is not, contrary to our supposition, which completes the proof that $y\star$ is the least upper bound of A. Q.E.D.

EXERCISES

1. Prove Theorems 44 and 45.
2. Prove Theorem 46. (This exercise has fifteen parts.)
3. Prove Theorems 47 and 48.

4. Prove Theorem 50.

5. Prove Theorem 51.

6. Prove Theorems 53 and 54.

7. Prove that an empty set of real numbers does not have a least upper bound.

8. Let x and y be sequences of real numbers which have a limit. Prove that

   (a) $\lim\limits_{n \to \infty} (x_n + y_n) = \lim\limits_{n \to \infty} x_n + \lim\limits_{n \to \infty} y_n,$

   (b) $\lim\limits_{n \to \infty} (x_n y_n) = \lim\limits_{n \to \infty} x_n \cdot \lim\limits_{n \to \infty} y_n,$

   (c) if for every n, $y_n \neq 0$ and $\lim\limits_{n \to \infty} y_n \neq 0$ then

$$\lim_{n \to \infty} \left(\frac{x_n}{y_n}\right) = \frac{\lim\limits_{n \to \infty} x_n}{\lim\limits_{n \to \infty} y_n}.$$

9. Let x be a Cauchy sequence of real numbers and let y be a subsequence of x. Prove that

$$\lim_{n \to \infty} y_n = \lim_{n \to \infty} x_n.$$

10. Define in the obvious manner the notions of lower bound and greatest lower bound of a set of real numbers, and prove that every non-empty set of real numbers which has a lower bound has a greatest lower bound.

11. A sequence x of real numbers is said to be *bounded* if there is a real number $x\star$ such that for every n

$$|x_n| \leq x\star.$$

Is every Cauchy sequence of real numbers bounded? If so, prove it. If not, give a counterexample.

§ 6.7 Sets of the Power of the Continuum.

The first business of this final section of the chapter is to prove Cantor's famous theorem of 1874 that the set of real numbers is not denumerable. A useful preliminary is the theorem that every real number can be uniquely represented by a non-terminating decimal. The exact formulation of the latter is the following, generalized to any integer radix ≥ 2.

THEOREM 56. *Let r be an integer ≥ 2. Every real number x is uniquely representable with respect to the radix r as a sequence $\langle a, d_1, d_2, \ldots, d_n, \ldots \rangle$ such that*

   (i) *a is the largest integer equal to or less than x,*

   (ii) *for all n, $0 \leq d_n < r$ and d_n is an integer,*

   (iii) *it is not the case that there is an N such that for all $n > N$, $d_n = r - 1$,*

   (iv) *the sequence whose terms c_n are defined recursively by*

$$c_0 = a$$
$$c_{n+1} = c_n + d_{n+1}/r^{n+1}$$

is a Cauchy sequence which converges to x.

PROOF. Let x be any real number. Let ar be the greatest integer which is equal to or less than rx. Then there is a non-negative number $\epsilon_1 < r$ such that

$$x = a + \frac{\epsilon_1}{r}.$$

In a similar manner

$$\epsilon_1 = d_1 + \frac{\epsilon_2}{r}, \quad \epsilon_2 = d_2 + \frac{\epsilon_3}{r}, \cdots, \quad \epsilon_n = d_n + \frac{\epsilon_{n+1}}{r},$$

where for all n, $0 \leq d_n$, $\epsilon_n < r$ and d_n is an integer. Thus

$$x = a + \frac{d_1}{r} + \frac{d_2}{r^2} + \cdots + \frac{d_n}{r^n} + \frac{\epsilon_{n+1}}{r^{n+1}},$$

and consequently

$$0 \leq x - \left(a + \frac{d_1}{r} + \frac{d_2}{r^2} + \cdots + \frac{d_n}{r^n} \right) < \frac{1}{r^n},$$

that is, for every n, using the definition of (iv),

$$0 \leq x - c_n < \frac{1}{r^n},$$

but $\lim\limits_{n \to \infty} \frac{1}{r_n} = 0$, whence $\lim\limits_{n \to \infty} c_n = x$. Q.E.D.

Note that condition (iii) of the theorem eliminates for the radix 10, for example, an infinite string of 9's. Thus 5.000 . . . cannot, following the theorem, also be represented by 4.999 . . . , which representation must be excluded to guarantee uniqueness.

Proof of the converse of Theorem 56 we leave as an exercise.

THEOREM 57. *Let $\langle a, d_1, d_2, \ldots, d_n, \ldots \rangle$ be a sequence such that*

(i)   *a is an integer,*

(ii) *there is an integer $r \geq 2$ such that for all n, $0 \leq d_n < r$ and every d_n is an integer.*

Then there is a unique real number x such that the sequence whose terms c_n are defined recursively by: $c_0 = a$ and $c_{n+1} = c_n + d_{n+1}/r^{n+1}$, is a Cauchy sequence which converges to x.

By choosing the radix 2 it is not difficult, on the basis of Theorems 56 and 57, to prove:

THEOREM 58. *The set of all real numbers is equipollent to the set 2^ω.*

By virtue of Theorems 14 and 23 of Chapter 4, it then follows from Theorem 58 that

THEOREM 59.   *The set of real numbers is not denumerable.*

It will also be useful to give Cantor's more constructive proof of this theorem by use of his important "diagonal method." We use Theorem 56 and represent the real numbers by their decimal expansions. Obviously the set of real numbers is infinite. Suppose now that it is also denumerable. Then there is a 1–1 function f from ω to the set of real numbers. Let

$$r_n = f(n).$$

Each real number r_n we represent in decimal notation as

$$r_n = a^{(n)} . d_1^{(n)} d_2^{(n)} d_3^{(n)} \ldots,$$

where $a^{(n)}$ is the largest integer equal to or less than r_n.
Now consider the real number

$$c = a . d_1 d_2 d_3 \ldots$$

defined as follows:

$$a = \begin{cases} 2 \text{ if } a^{(0)} = 1 \\ \\ 1 \text{ if } a^{(0)} \neq 1. \end{cases}$$

(Here $a^{(0)}$ is, of course, the largest integer equal to or less than r_0.) And

$$d_n = \begin{cases} 2 \text{ if } d^{(n)} = 1 \\ \\ 1 \text{ if } d^{(n)} \neq 1. \end{cases}$$

Thus if,

$$r_0 = 4.333\ldots$$
$$r_1 = 7.12171217\ldots$$
$$r_2 = 0.689689\ldots$$
$$r_3 = 0.414141\ldots,$$

then

$$c = 1.211\ldots.$$

But there can be no r_n such that

$$c = r_n$$

for the n-th decimal of r_n must differ from the n-th decimal of c. (We count the integer a as the 0-th decimal.) Hence there is no n such that

$$c = f(n),$$

and our supposition is false. Whence the set of real numbers is not denumerable.

Sets equipollent to the set of real numbers are often called *sets of the power of the continuum*. Some basic facts about such sets are summarized in the following theorems, which we do not prove in detail.

> **Theorem 60.** *The union of a finite or denumerable set with a set of the power of the continuum is again a set of the power of the continuum.*

proof. Compare the proof of Theorem 67 of Chapter 5.

> **Theorem 61.** *If A is a set of the power of the continuum and B is a finite or denumerable set then $A \sim B$ is a set of the power of the continuum.*

> **Theorem 62.** *The set of all real numbers constituting the interval between any two distinct real numbers a and b is of the power of the continuum.*

proof. Consider first the mapping f of the positive real numbers onto the interval $(0,1)$:

$$f(x) = \frac{x}{1+x}$$

In a similar fashion map the negative real numbers onto the interval $(-1,0)$. It follows at once from these two 1–1 mappings that the set of real numbers in the interval $(-1,1)$ is of the power of the continuum. We leave it as an exercise to construct a 1–1 function mapping any finite interval (a,b) onto $(-1,1)$.

> **Theorem 63.** *Every set of the power of the continuum may be represented as the union of n mutually exclusive subsets each of which is of the power of the continuum.*

The proof is by induction on n.
Moreover,

> **Theorem 64.** *Every set of the power of the continuum may be represented as the union of an infinite sequence of sets without common elements and each of which is of the power of the continuum.*

proof. Use the sequence $\langle A_1, A_2, \ldots, A_n, \ldots \rangle$ defined by: A_n is the set of all real x numbers such that $n - 1 \leq x < n$.

By the same approach it is easy to show that

> **Theorem 65.** *The Cartesian product of two sets of which one is denumerable and the other is of the power of the continuum is of the power of the continuum.*

We may also show

THEOREM 66. *The Cartesian product of two sets of the power of the continuum is also of the power of the continuum.*

PROOF. We show that $(0,1) \times (0,1) \approx (0,1)$. For this purpose, similar to the decimal representation of Theorem 56, we may represent any real number x in the interval $(0,1)$ as

$$x = \frac{d_1}{2} + \frac{d_2}{4} + \frac{d_3}{8} + \ldots, \quad d_n = 0 \text{ or } 1.$$

Considering only $d_n = 1$, we have

$$x = (\tfrac{1}{2})^{p_1} + (\tfrac{1}{2})^{p_2} + (\tfrac{1}{2})^{p_3} + \ldots,$$

with $p_1 < p_2 < p_3 < \ldots$. Define now

$$a_1 = p_1, \ a_2 = p_2 - p_1, \ a_3 = p_3 - p_2, \ \ldots$$

and represent

$$x = (\tfrac{1}{2})^{a_1} + (\tfrac{1}{2})^{a_1 + a_2} + (\tfrac{1}{2})^{a_1 + a_2 + a_3} + \ldots,$$

or in simpler notation we may represent x by the sequence $< a_1, a_2, a_3 \ldots, a_n, \ldots >$. We have a similar representation of $y \epsilon (0,1)$ as $< b_1, b_2, b_3, \ldots, b_n, \ldots >$. We define then $f(< x,y >)$ as the number in $(0,1)$ represented by the sequence $< a_1, b_1, a_2, b_2, a_3, b_3, \ldots, a_n, b_n, \ldots >$. It is straightforward to show that f is a 1–1 mapping of $(0,1) \times (0,1)$ onto $(0,1)$.*

Using the special axiom for cardinals, we may define

†DEFINITION 63. $\mathfrak{c} = \mathcal{K}(\text{Re})$.

The symbol 'c' is the standard one for the cardinality of the continuum. Using this definition, we state a general theorem on the cardinal arithmetic of $\mathfrak{c}$. Proof of each part of the theorem follows directly from one or more of the immediately preceding theorems of this section.

†THEOREM 67.

(i)   *If $\mathfrak{n}$ is a finite cardinal then*

$$\mathfrak{n} + \mathfrak{c} = \mathfrak{n} \cdot \mathfrak{c} = \mathfrak{c},$$

(ii) $\aleph_0 + \mathfrak{c} = \mathfrak{c}$,

(iii) $\aleph_0 \cdot \mathfrak{c} = \mathfrak{c}$,

(iv) $2^{\aleph_0} = \mathfrak{c}$.

(v) $\mathfrak{c} + \mathfrak{c} = \mathfrak{c}$.

(vi) $\mathfrak{c} \cdot \mathfrak{c} = \mathfrak{c}$.

It is not known whether there exists a set which is of greater power than a denumerable set and of less power than the continuum. The conjecture that there are no such sets is called the *Continuum Hypothesis*. Sierpinski [1956] summarizes in thorough fashion the known important consequences of this hypothesis. One of the most interesting, which is actually equivalent

---

*The proof is not too difficult that there is no 1–1 continuous function which will establish this correspondence (see Sierpinski [1958, p. 66]).

to the Continuum Hypothesis, is the following proposition (the equivalence was proved by Sierpinski in 1919): The set of all the points of the plane is the union of two sets of which one is at most denumerable along any line parallel to the ordinate axis and the other is at most denumerable along any line parallel to the abscissa axis.

We showed earlier that the set of all real numbers is equipollent to 2^ω. This relationship is the basis for the *Generalized Continuum Hypothesis*: given any infinite set A there is no set of power greater than A and less than the set of all subsets of A. Gödel ([1938], [1940]) has proved the important result that the Generalized Continuum Hypothesis may be consistently added to the other axioms of set theory,* that is, if the other axioms are jointly consistent then the addition of the Generalized Continuum Hypothesis as a new axiom will not give rise to a contradiction.

EXERCISES

1. Prove Theorem 57.
2. Prove Theorem 58.
3. Prove Theorem 60.
4. Prove Theorem 61.
5. Complete the proof of Theorem 62.
6. Prove Theorem 63.
7. Prove Theorem 64.
8. Prove Theorem 65.
9. Prove Theorem 66.
10. Prove that there is no continuous mapping of the plane onto the straight line.
11. Prove Theorem 67.

---

*His proof is actually for the axioms of von Neumann set theory, as given in his papers, but with little modification the proof holds for the axioms of Zermelo-Fraenkel set theory as well.

CHAPTER 7

TRANSFINITE INDUCTION AND ORDINAL ARITHMETIC

§ 7.1 Transfinite Induction and Definition by Transfinite Recursion. The focus of the present chapter is ordinal number theory. The first section considers transfinite induction and definition by transfinite recursion for ordinal numbers. The axiom schema of replacement is needed to justify the general recursion schema. The second section presents the elements of ordinal arithmetic. The third section considers cardinal numbers again but without use of the special axiom for cardinals. In particular the concept of an aleph is introduced. The fourth section generalizes the results of §7.1 to well-ordered sets. In addition, the fundamental theorem for well-ordered sets is proved. The final section (§7.5) presents a revised summary of the axioms; the axiom schema of replacement is used to derive the pairing axiom and the axiom schema of separation.

Our first problem is to extend the principle of induction for natural numbers (finite ordinals) to the principle of transfinite induction for all ordinals. As Theorem 12 of §5.1 establishes, the set of all ordinals does not exist, so that a general set formulation of transfinite induction cannot be given. However, as we shall see, a set formulation up to any given ordinal is possible. Also, a general schema for all ordinals analogous to Theorem 22 of §5.2 can be proved; in fact, we begin with this.

The initial letters of the Greek alphabet in lower-case type, that is, 'α', 'β', 'γ', 'δ', . . . , with and without subscripts, will indicate variables which take ordinals as values.

THEOREM SCHEMA 1. [Principle of Transfinite Induction: First Formulation]. *If for every α*

(i) $(\forall \beta)(\beta < \alpha \rightarrow \varphi(\beta)) \rightarrow \varphi(\alpha)$

then for every α, $\varphi(\alpha)$.

PROOF. Granted the hypothesis of the theorem, suppose there is an α such that $\varphi(\alpha)$ is false. Let

$$L(\alpha) = \{\beta : \beta \leq \alpha \ \& \ \varphi(\beta) \text{ is false}\}.$$

Then $\mathcal{E}\alpha^\mathsf{I}$ well-orders $L(\alpha)$. Let β^* be the first element of $L(\alpha)$. By hypothesis on β^* for every $\gamma < \beta^*$ we have: $\varphi(\gamma)$. But then by the (inductive) hypothesis, that is, (i) of the theorem, $\varphi(\beta^*)$, which is a contradiction. Q.E.D.

The formulation of Theorem 1, when specialized to the natural numbers, yields what is often called a "course-of-values" induction, because *all* natural numbers less than a given number are considered, not just the immediate predecessor.

The proof of the set formulation of transfinite induction is a trivial variant of the above, so we only state the theorem, but the fact that it, in contrast to Theorem 1, is relative to an ordinal α should be noted.

THEOREM 2. [Principle of Transfinite Induction: Second Formulation].
If for every ordinal $\beta < \alpha$ we have that $\beta \subseteq A$ implies that $\beta \in A$, then $\alpha \subseteq A$. Symbolically:

$$(\forall\beta)(\beta < \alpha \ \& \ \beta \subseteq A \rightarrow \beta \in A) \rightarrow \alpha \subseteq A.$$

If we take $\alpha = \omega$, we have as a special case of the theorem a formulation of induction for the natural numbers slightly different from Theorem 24 of §5.2:

If $m \in A$ whenever $m \subseteq A$ then $\omega \subseteq A$.

The formulation of transfinite induction provided by Theorem 2 has a certain interest from the standpoint of applications. In using transfinite induction to prove a theorem, it is often desired to know how "far up" the induction goes, that is, what is the smallest ordinal α which will suffice in applying Theorem 2. Of course, in ordinary induction we always take $\alpha = \omega$.

We now want to state without proof a third formulation of transfinite induction which uses a condition like the standard one for induction on the natural numbers: if $\varphi(n)$ then $\varphi(n^\mathsf{I})$, and restricts the full course-of-values inductions to *limit ordinals*, that is, ordinals which have no immediate predecessor and are not zero.

DEFINITION 1. *α is a limit ordinal if and only if $\alpha \neq 0$ and there is no β such that $\beta^\mathsf{I} = \alpha$.*

The ordinal ω is the only limit ordinal we have introduced so far. A useful fact about limit ordinals is stated here without proof.

THEOREM 3. *If α is a limit ordinal then $\bigcup\alpha = \alpha$.*

THEOREM SCHEMA 4. [Principle of Transfinite Induction: Third Formulation].

Suppose that

(i) $\varphi(0)$,

(ii) *for every α, if $\varphi(\alpha)$ then $\varphi(\alpha^{\mathbf{I}})$, and*

(iii) *for every limit ordinal γ, if for every $\beta < \gamma$, $\varphi(\beta)$, then $\varphi(\gamma)$.*

Then for every ordinal α, $\varphi(\alpha)$.

Before turning to transfinite recursions, a useful theorem schema on the existence of a least ordinal satisfying a given property may be stated. Its proof is left as an exercise.

THEOREM SCHEMA 5. *If there is an α such that $\varphi(\alpha)$ then there is a least ordinal β such that $\varphi(\beta)$.*

In order subsequently to introduce the ordinal operations of addition, multiplication, and exponentiation, it is not sufficient to have the principle of transfinite induction at hand. It is necessary to justify definition by transfinite recursion in the same general way that we justified definition by recursion of the operations on natural numbers in §5.2. As far as I know, the first proof of a general theorem justifying definition by transfinite recursion is to be found in von Neumann [1928b].

The proof of the first formulation parallels very closely the proof of Theorem 27 of §5.2 on definition by recursion. The major difference in formulation is that the function H (the theorem says 'set H' but in applications we are always interested in functions) takes the whole of F restricted to β as its argument, rather than simply the predecessor as in Theorem 27. An unadulterated form of Theorem 27 could not, of course, be used for ordinals in general, since limit ordinals have no immediate predecessors.

THEOREM 6. [Transfinite Recursion: First Formulation]. *Let H be any set and α any ordinal. Then there exists a unique F such that*

(i)   *F is a function on α,*

(ii) *for every $\beta < \alpha$*

$$F(\beta) = H(F \mid \beta).$$

PROOF. First by virtue of the axiom schema of separation there is a set A such that $f \in A$ if and only if:

(1) $f \in \mathcal{P}(\alpha \times (\mathcal{R}H \cup \{0\}))$,

and there is a β such that

(2) f is a function on β,

(3)   for every γ, if $\gamma < \beta$ then $f(\gamma) = H(f \mid \gamma)$.

As in the case of Theorem 27 of §5.2 the idea of the proof is to show that $\cup A$ is the desired function F. We begin by showing:

(4)   If f, $g \in A$ then $f \subseteq g$ or $g \subseteq f$.

Let β_1 be the domain of f and β_2 the domain of g. Then $\beta_1 \cap \beta_2$ is β_1 or β_2. Now suppose there is a γ in $\beta_1 \cap \beta_2$ such that

$$f(\gamma) \neq g(\gamma).$$

Let $\gamma \star$ be the smallest such ordinal (see Theorem 5). We then have:

$$f \mid \gamma \star = g \mid \gamma \star,$$

and *a fortiori*

(5) $$H(f \mid \gamma \star) = H(g \mid \gamma \star),$$

but then

$$f(\gamma \star) = g(\gamma \star),$$

contrary to our supposition, whence (4) is established.

Now we define:

$$F = \cup A.$$

It follows at once from (4) that F is a function. Furthermore if β is in the domain of F, then for some f in A, β is in the domain of f, and consequently

$$f(\beta) = H(f \mid \beta),$$

whence

$$F(\beta) = H(F \mid \beta).$$

It only remains to show that α is the domain of F. Suppose not. Let β^* be the smallest ordinal less than α and not in the domain of F. Then there is an f in A whose domain is β^*. Hence

$$f \cup [\langle \beta^*, H(f \mid \beta^*) \rangle] \in A,$$

and therefore $\beta \star$ is in the domain of F, contrary to our supposition. Proof that F is unique is left as an exercise. Q.E.D.

As might be expected, by making use of limit ordinals a version of transfinite recursion may be given which is closer to Theorem 27 of §5.2. The intuitive content of condition (iv) of the theorem will be discussed in relation to the definitions of ordinal addition and multiplication in §7.2.

THEOREM. 7.   [Transfinite Recursion: Second Formulation].   *Let x be any object, G any set, and α any non-zero ordinal. Then there is a unique F such that*

(i)   *F is a function on α,*

(ii) $F(0) = x$,

(iii)   *for every β with $\beta^1 < \alpha$,*

$$F(\beta^1) = G(F(\beta)),$$

(iv)   *for every $\beta < \alpha$ if β is a limit ordinal then*

$$F(\beta) = \bigcup_{\gamma \in \beta} F(\gamma).$$

PROOF. The proof of this formulation of the theorem justifying transfinite recursion is got by specifying the appropriate function H in the preceding theorem.

Let

$$L = \mathcal{P}(\cup(\mathcal{R}G \cup \{x\})) \cup \mathcal{R}G \cup \{x\}.$$

We first define the notion of an α-sequence of L. An α-sequence of L is a function whose domain is α and whose range is a subset of L. (A sequence, in the usual mathematical sense, of elements of L is just an ω-sequence.) Now consider the function H defined as follows:

(1)   The domain of H is the set of β-sequences of L with $\beta < \alpha$,

(2)   For the 0-sequence 0

$$H(0) = x,$$

(3)   For any β^1-sequence s

$$H(s) = G(s(\beta)),$$

(4)   For β a limit ordinal and s a β-sequence

$$H(s) = \bigcup_{\gamma \in \beta} s(\gamma).$$

Note that by virtue of Definition 15 of §2.6

$$\bigcup_{\gamma \in \beta} s(\gamma) = \cup \{y : (\exists \gamma)(\gamma \in \beta \,\&\, y = s(\gamma))\},$$

and because $s(\gamma) \in L$, it follows from the axiom schema of separation that the set $\{y : (\exists \gamma)(\gamma \in \beta \,\&\, y = s(\gamma))\}$ always exists in the appropriate sense, i.e., is not empty. (Concerning the range of H, see Exercise 6 below.) In view of the preceding theorem (Theorem 6) there is a unique function F defined on α such that for every $\beta < \alpha$

(5) $$F(\beta) = H(F \mid \beta).$$

We observe that $F \mid \beta$ is a β-sequence. Whence by (2) and (5)

$$F(0) = H(F \mid 0) = x,$$

establishing (ii) of the theorem.

On the basis of (3) and (5)

$$F(\beta\prime) = H(F \mid \beta\prime) = G([F \mid \beta\prime](\beta)) = G(F(\beta)),$$

which proves (iii).

Finally, in view of (4) and (5), if β is a limit ordinal,

$$F(\beta) = H(F \mid \beta) = \bigcup_{\gamma \in \beta} ([F \mid \beta](\gamma)) = \bigcup_{\gamma \in \beta} (F(\gamma)).$$

Uniqueness of F is obvious. Q.E.D.

It should be noted that Theorem 7 is less general than Theorem 6, because not all the functions F from Theorem 6 satisfy (iv) of Theorem 7, that is, satisfy:

$$F(\beta) = \bigcup_{\gamma \in \beta} F(\gamma)$$

for β a limit ordinal. For example, in Theorem 6, let α be $\omega\prime$ and for any β-sequence s with $\beta \leq \omega\prime$ define

$$H(s) = \begin{cases} 0 \text{ if } \beta \text{ is a limit ordinal} \\ \\ 1 \text{ otherwise.} \end{cases}$$

Then

$$F(\omega) = 0,$$

but

$$\bigcup_{n \in \omega} F(n) = \bigcup \{1\} = 1.$$

It should be clear why Theorems 6 and 7 do not provide the optimal justification for introduction of the operations of ordinal addition. The set of all ordinals does not exist and thus it is hopeless to define ordinal addition or multiplication as a set-theoretical function. The remaining objective of this section is to establish a theorem schema which will justify definition of the appropriate operation symbols. To begin with, we prove a transfinite recursion schema analogous to Theorem 6. This schema has been placed after Theorems 6 and 7 because its proof requires the axiom schema of replacement, which we introduce below.

The need for a schematic formulation of transfinite recursion may perhaps be made more apparent by discussing the difficulties which arise when we attempt to use Theorem 7 to define ordinal addition. The recursive scheme we have in mind is simply:

(i) $\alpha + 0 = \alpha$,

(ii) $\alpha + \beta' = (\alpha + \beta)'$

(iii)   If β is a limit ordinal

$$\alpha + \beta = \bigcup_{\gamma \in \beta} (\alpha + \gamma).$$

Conditions (i) and (ii) yield a recursive scheme like that of integer addition. The third condition is new. We may illustrate its intuitive meaning by considering $\omega + \omega$. We have not yet proved that $\omega + \omega$ exists, but leaving that question aside for the moment, the basic idea is that

$$\omega + \omega = \{0, 1, 2, 3, \ldots, \omega, \omega + 1, \omega + 2, \omega + 3, \ldots\}.$$

And we easily see that this is just what we get from:

$$\omega + \omega = \bigcup_{\gamma \in \omega} (\omega + \gamma) = \bigcup_{n \in \omega} (\omega + n).$$

The form of (iii) should make clear why (iv) of Theorem 7 has the form it does.

Unfortunately, complications immediately arise if we attempt to apply Theorem 7 to convert these three conditions into a precise definition of a unary function $\alpha+$. (Because Theorem 7 yields only unary functions, for each ordinal α, we define $\alpha+$, rather than the binary operation symbol $+$.) First, we cannot simply define a function $\alpha+$ for all ordinals, since the set of all ordinals does not exist. Suppose then we define $\alpha+$ with respect to some ordinal η, which requires a subscript: $\alpha +_\eta$. Then we would have:

(1)
$$\begin{cases} \alpha +_\eta 0 = \alpha \\ \alpha +_\eta \beta' = (\alpha +_\eta \beta)' \end{cases}$$

and so forth, and we would naturally impose the restriction that the domain of $\alpha+$ is η. However, the relativization of ordinal addition to some ordinal η is not sufficient to permit an application of Theorem 7. As remarked prior to the definition of integer addition, we cannot simply use the successor symbol in picking a particular function G, for there is no set designated by the successor symbol. To deal with integer addition we introduced the notation '$\mathfrak{S}_A$' to designate the successor function restricted to A. Probably the natural suggestion is then to try $\mathfrak{S}\eta$ in (1):

(2)
$$\begin{cases} \alpha +_\eta 0 = \alpha \\ \alpha +_\eta \mathfrak{S}_\eta(\beta) = \mathfrak{S}_\eta(\alpha +_\eta \beta). \end{cases}$$

There is no difficulty about $\mathfrak{S}_\eta(\beta)$ since $\beta \in \eta$, but the situation is different for $\mathfrak{S}_\eta(\alpha +_\eta \beta)$. If α is big enough it can happen that $\alpha +_\eta \beta > \eta$, and therefore $\mathfrak{S}_\eta(\alpha +_\eta \beta) = 0$. For example,

(3) $$\mathfrak{S}_\omega(\omega +_\omega 3) = 0,$$

whereas,

(4) $$(\omega +_\omega 3)^1 = \omega +_\omega 4 \neq 0.$$

The difficulty raised by (2) and the particular case (3) is fundamental, and on the basis of Theorem 7 it is impossible to make repairs which will permit construction of a reasonable ordinal arithmetic.

As already remarked, to prove the appropriate recursion theorem we need the axiom schema of replacement. The intuitive idea of the axiom is that if we have a formula $\varphi(x,y)$ with the functional property that for every x in a set A there is at most one y such that $\varphi(x,y)$, then we may assert that the *set* of y's exists, and "replace" A by this new set. Formally, we have:*

AXIOM SCHEMA OF REPLACEMENT. *If*

$$(\forall x)(\forall y)(\forall z)(x \in A \ \& \ \varphi(x,y) \ \& \ \varphi(x,z) \to y = z)$$

then

$$(\exists B)(\forall y)(y \in B \leftrightarrow (\exists x)(x \in A \ \& \ \varphi(x,y))).$$

The hypothesis of the axiom simply requires that $\varphi(x,y)$ be functional in x. If this requirement were dropped we would immediately be in difficulty. For example, we could let $\varphi(x,y)$ be

$$x \subseteq y,$$

and let $A = \{0\}$, then because the empty set is a subset of every set, B would be the paradoxical set of all sets.

In the last section of this chapter we show that the axiom schema of separation follows from the axiom schema of replacement (the proof is trivial) and also that the pairing axiom follows from the power set axiom and the axiom schema of replacement.

THEOREM SCHEMA 8. [Transfinite Recursion: Third Formulation].†

Let τ be any term. Then the following is a theorem: For any ordinal α there is a unique F such that

---

*This axiom is also called the *axiom schema of substitution;* it originates with Fraenkel [1922]. In the literature of set theory the axiom is usually credited solely to Fraenkel, but Skolem [1922, p. 226] formulated it independently and at the same time. In Fraenkel [1928], it may be said, Skolem's independent formulation is acknowledged.

†This formulation is rather close to the original one in von Neumann [1928].

(i) F is a function on α,
(ii) for every $\beta < \alpha$

$$F(\beta) = \tau(F \mid \beta).$$

PROOF. Paralleling now the first step in the proof of Theorem 6, we take $\varphi(\beta,f)$ to be:

$$\varphi(\beta,f) \leftrightarrow f \text{ is a function on } \beta \ \& \ (\boldsymbol{\forall}\gamma)(\gamma < \beta \rightarrow f(\gamma) = \tau(f\mid\gamma)).$$

If we have: $\varphi(\beta,f) \ \& \ \varphi(\beta,f')$, it easily follows by transfinite induction (Theorem 1) that $f = f'$, for suppose there were a γ such that $f(\gamma) \neq f'(\gamma)$. Let $\gamma \star$ be the first such ordinal. Then

$$f \mid \gamma \star = f' \mid \gamma \star,$$

and a fortiori

$$\tau(f \mid \gamma \star) = \tau(f' \mid \gamma \star),$$

whence

$$f(\gamma \star) = f'(\gamma \star),$$

which is absurd.

Since φ has the proper many-one property, we may apply the axiom schema of replacement to obtain: there is a set A such that

$$f \in A \leftrightarrow (\exists\beta)(\beta \in \alpha^{\textsf{I}} \ \& \ \varphi(\beta,f)).$$

The remainder of the proof simply consists in showing that $\bigcup A$ is the desired function F, and is essentially the same as the corresponding part of Theorem 6. Q.E.D.

Analogous to Theorem 7 we have another formulation in the form of a schema. We leave the proof as an exercise.

THEOREM SCHEMA 9. [Transfinite Recursion: Fourth Formulation]. Let τ be any term. Then the following is a theorem: If x is any object and α any non-zero ordinal, then there is a unique F such that

(i) F is a function on α,
(ii) $F(0) = x$,
(iii) for every ordinal β with $\beta^{\textsf{I}} < \alpha$,

$$F(\beta^{\textsf{I}}) = \tau(F(\beta)),$$

(iv) for every limit ordinal $\beta < \alpha$

$$F(\beta) = \bigcup_{\gamma \in \beta} F(\gamma).$$

For the purposes of applications it is necessary to know that a function defined by transfinite recursion (via Theorems 6-9) is independent of the particular ordinal chosen in the sense expressed by the next theorem. The proof is left as an exercise. It is understood in the formulation of the theorem that whichever one of the four theorems is used, the functions F_1 and F_2 are defined with respect to the same object x, set F, or term τ, as the case may be.

THEOREM 10. *Let F_1 and F_2 be defined by transfinite recursion up to α_1 and α_2 respectively by use of Theorem 6, 7, 8, or 9. Then*

$$F_1 \mid \alpha_2 = F_2 \mid \alpha_1.$$

On the basis of Theorem 9 we may define ordinal addition as follows:

$\alpha + \beta = \gamma$ *if and only if there is a function f such that*

(i)   *f is a function on β^1,*

(ii) *$f(0) = \alpha$,*

(iii) *for every η with $\eta < \beta$*

$$f(\eta^1) = f(\eta)^1,$$

(iv) *for every limit ordinal $\eta < \beta^1$*

$$f(\eta) = \bigcup_{\theta \subset \eta} f(\theta),$$

(v)   *$f(\beta) = \gamma$.*

In this definition we have essentially defined a hierarchy of one-place functions $\alpha +$, for each α with domain β^1. An esthetically more satisfying procedure is to use Theorems 9 and 10 to prove a general schema on the basis of which a more intuitive definition of binary ordinal operations like addition may be given. For this general schema we need the following general result, whose proof depends on the axiom schema of replacement.

THEOREM SCHEMA 11.

$$y \in \bigcup_{x \in A} \tau(x) \leftrightarrow (\exists x)(\exists B)(x \in A \,\&\, B = \tau(x) \,\&\, y \in B)$$

We may then establish on the basis of preceding theorems:

THEOREM SCHEMA 12. [Transfinite Recursion: Fifth Formulation]. *Let $\sigma(\alpha_1, \ldots, \alpha_{n-1})$ be any term with at most $n - 1$ free variables, and let $\mu(\alpha_1, \ldots, \alpha_n)$ be any term with at most n free variables. Then a term $\tau(\alpha_1, \ldots, \alpha_n)$ may be defined by the recursion schema:*

(i)   *$\tau(\alpha_1, \ldots, \alpha_{n-1}, 0) = \sigma(\alpha_1, \ldots, \alpha_{n-1})$,*

(ii) *for every β*
$$\tau(\alpha_1, \ldots, \alpha_{n-1}, \beta^1) = \mu(\alpha_1, \ldots, \alpha_{n-1}, \tau(\alpha_1, \ldots, \alpha_{n-1}, \beta)),$$

(iii) *for every limit ordinal β*

$$\tau(\alpha_1, \ldots, \alpha_{n-1}, \beta) = \bigcup_{\gamma \in \beta} \tau(\alpha_1, \ldots, \alpha_{n-1}, \gamma).$$

The free variables $\alpha_1, \ldots, \alpha_{n-1}$ are the parameters of recursion in this theorem. Given this theorem, we may in §7.2 use the appropriate recursion schemata themselves to define ordinal addition, multiplication, and exponentiation.

EXERCISES

1. Prove Theorem 3.
2. Prove Theorem 4.
3. Prove Theorem 5.
4. Complete the proof of Theorem 6 by showing that F is unique.
5. Give another example than the one in the text of a function F from Theorem 6 which does not satisfy

$$F(\beta) = \bigcup_{\gamma \in \beta} F(\gamma)$$

for β a limit ordinal.

6. In the proof of Theorem 7, we need to know that the range of H is a set in order to know that H as a function is a well-defined set. Of what set is the range of H a subset?

7. Prove Theorem 9.
8. Prove Theorem 10.
9. Prove Theorem 11.
10. Prove Theorem 12.

§ 7.2 Elements of Ordinal Arithmetic.

Our intention is to give the main outlines of ordinal arithmetic, but not to develop it in any complete sense (in this connection see Sierpinski [1928] and Bachmann [1955]).

The initial theorems on ordinal addition and multiplication parallel those on integer addition and multiplication in §5.2. Proofs by transfinite induction here play the role that proofs by induction played in §5.2. Considered as generalizations of the integer operations, the most striking single fact about ordinal addition and multiplication is that they are not commutative. In contrast, cardinal addition and multiplication are commutative, as we saw in §4.3.

We may use Theorem 12 to introduce ordinal addition by the recursion scheme already given in §7.1.

DEFINITION 2. *The operation of ordinal addition is defined by the following recursion scheme:*

(i) $\alpha + 0 = \alpha,$
(ii) $\alpha + \beta^\mathsf{I} = (\alpha + \beta)^\mathsf{I},$
(iii) *if β is a limit ordinal*

$$\alpha + \beta = \bigcup_{\gamma \in \beta} (\alpha + \gamma).$$

The existence of $\underset{\gamma \in \beta}{\bigcup} (\alpha + \gamma)$ in the appropriate intuitive-sense follows from earlier results in §7.1. The role of (iii) will become clear in subsequent proofs of theorems in this section.

We leave as an exercise proof of the following result which we use recurrently without explicit mention.

THEOREM 13. $\alpha + \beta$ *is an ordinal.*

We turn now to the elementary arithmetic of ordinal addition. Many of the proofs exhibit typical methods of argument by transfinite induction.

THEOREM 14. $\alpha + 0 = 0 + \alpha = \alpha$.

PROOF. By (i) of the definition we have immediately that

$$\alpha + 0 = \alpha.$$

The proof that $0 + \alpha = \alpha$ is more complicated; it exhibits the typical three parts in a direct argument by transfinite induction (as formulated by Theorem 4).

Part 1. $\alpha = 0$. We need to show:

$$0 + 0 = 0.$$

Part 2. Assuming $0 + \alpha = \alpha$, we need to show:

$$0 + \alpha' = \alpha'.$$

Part 3. Assuming that α is a limit ordinal and that for $\beta < \alpha$,

$$0 + \beta = \beta,$$

we need to show:

$$0 + \alpha = \alpha.$$

Part 1 follows at once from (i) of the definition of addition.

To establish Part 2, we use (ii) of the definition and our inductive hypothesis:

$$0 + \alpha' = (0 + \alpha)'$$
$$= \alpha'.$$

To prove Part 3 we use (iii) of the definition:

$$0 + \alpha = \underset{\beta \in \alpha}{\bigcup} (0 + \beta)$$
$$= \underset{\beta \in \alpha}{\bigcup} \beta \qquad \text{by inductive hypothesis}$$
$$= \alpha \qquad \text{by Theorem 3}$$

since if α is a limit ordinal, $\bigcup \alpha = \alpha$.

Having proved Parts 1-3, we conclude immediately by transfinite induction (Theorem 4) that our theorem holds for all ordinals. Q.E.D.

The next theorem is an immediate consequence of the definition of ordinal addition.

THEOREM 15. $\alpha + \beta^{\mathfrak{l}} = (\alpha + \beta)^{\mathfrak{l}}$.

Putting $\beta = 0$, we have:

THEOREM 16. $\alpha + 1 = \alpha^{\mathfrak{l}}$.

The following theorem establishes that ordinal addition is *right-monotonic* with respect to the relation *less than*.

THEOREM 17.   *If $\beta < \gamma$ then $\alpha + \beta < \alpha + \gamma$.*

PROOF.   The proof is by transfinite induction on γ.

Part 1.   If $\gamma = 0$ the hypothesis of the theorem is false and thus the theorem holds vacuously.

Part 2.   Suppose we have that if $\beta < \gamma$ then

$$\alpha + \beta < \alpha + \gamma.$$

To prove the theorem holds for the successor of γ we have two cases to consider.

Case 1. $\beta < \gamma$.   Then we have:

$$\alpha + \beta < \alpha + \gamma,$$

but

$$\alpha + \gamma < (\alpha + \gamma)^{\mathfrak{l}}$$

and by Theorem 15

$$(\alpha + \gamma)^{\mathfrak{l}} = \alpha + \gamma^{\mathfrak{l}},$$

whence by transitivity

$$\alpha + \beta < \alpha + \gamma^{\mathfrak{l}}.$$

Case 2. $\beta = \gamma$.   Then we have both

$$\beta < \gamma^{\mathfrak{l}}$$

and

$$\alpha + \beta = \alpha + \gamma,$$

but, as before

$$\alpha + \gamma < \alpha + \gamma^{\mathfrak{l}},$$

hence

$$\alpha + \beta < \alpha + \gamma^{\textsf{l}}.$$

(Note that we do not need to consider the case of $\gamma < \beta$, for by Theorem 15 of §5.1 we then have $\gamma^{\textsf{l}} \leq \beta$ and the theorem holds vacuously.)

Part 3. Suppose now that γ is a limit ordinal (this is the only part of the inductive hypothesis for this part that we need). Since $\beta < \gamma$, of course $\beta \in \gamma$. But by (iii) of the definition of addition

$$\alpha + \gamma = \bigcup_{\delta \in \gamma} (\alpha + \delta),$$

and so, in view of Definition 15 of §2.6 and Theorem 13,

$$\alpha + \gamma = \bigcup A$$

where

$$A = \{\eta \colon \exists \delta \in \gamma \,\&\, \eta = \alpha + \delta\}.$$

Now $\beta \in \gamma$, whence $\beta^{\textsf{l}} \in \gamma$ (for γ is a limit ordinal) and thus

$$\alpha + \beta^{\textsf{l}} \in A$$

but

$$(\alpha + \beta)^{\textsf{l}} = \alpha + \beta^{\textsf{l}}$$

and

$$\alpha + \beta \in (\alpha + \beta)^{\textsf{l}},$$

whence

$$\alpha + \beta \in \bigcup A,$$

i.e.,

$$\alpha + \beta \in \alpha + \gamma. \qquad\qquad \text{Q.E.D.}$$

On the other hand, ordinal addition is not left-monotonic with respect to *less than*, that is, we do not have in general

$$\text{if } \alpha < \beta \text{ then } \alpha + \gamma < \beta + \gamma.$$

For instance,

$$1 < 2,$$

but

$$1 + \omega = 2 + \omega.$$

(The verification of this fact we leave as an exercise.)

As an immediate consequence of the preceding theorem and Theorem 14, we have:

THEOREM 18. *If $\beta > 0$ then $\alpha + \beta > \alpha$.*

We also may easily derive a left cancellation law from the right-monotonicity of ordinal addition. The example given before Theorem 18 shows that cancellation from the right does not hold.

THEOREM 19. *If $\alpha + \beta = \alpha + \gamma$ then $\beta = \gamma$.*

PROOF. Suppose $\beta \neq \gamma$. For definiteness, let $\beta < \gamma$. Then by Theorem 17

$$\alpha + \beta < \alpha + \gamma,$$

which contradicts the hypothesis of the theorem. Q.E.D.

We do get a left-monotonic law for ordinal addition with respect to $\leq$. The proof is similar to that of Theorem 17 and is left as an exercise.

THEOREM 20. *If $\alpha \leq \beta$ then $\alpha + \gamma \leq \beta + \gamma$.*

As an immediate consequence we have:

THEOREM 21. $\alpha + \beta \geq \beta$.

We now prove a useful fact about limit ordinals.

THEOREM 22. *If β is a limit ordinal then $\alpha + \beta$ is a limit ordinal.*

PROOF. Suppose that $\alpha + \beta$ is not a limit ordinal. By virtue of Theorem 21 $\alpha + \beta \neq 0$. Hence there is a γ such that

$$(1) \qquad \gamma' = \alpha + \beta,$$
$$= \bigcup_{\delta \in \beta} (\alpha + \delta),$$

by virtue of (iii) of the definition of addition. Then

$$\gamma \in \bigcup_{\delta \in \beta} (\alpha + \delta),$$

hence there is a $\delta_1 \in \beta$ such that

$$\gamma \in \alpha + \delta_1$$

and consequently

$$(2) \qquad \gamma' < (\alpha + \delta_1)' = \alpha + \delta_1'.$$

(1) and (2) yield:

$$\alpha + \beta < \alpha + \delta_1'$$

and it follows from the contrapositive of Theorem 20 that

$$\beta < \delta_1',$$

which is absurd, because $\delta_1 \in \beta$ and thus $\delta_1' \in \beta$. Q.E.D.

The next theorem asserts that a central property of integer addition also holds for general ordinal addition.

THEOREM 23.   *If $\alpha \leq \beta$ then there is a unique γ such that $\alpha + \gamma = \beta$.*

PROOF. If $\alpha = \beta$, we take $\gamma = 0$. Suppose then that $\alpha < \beta$. Now by Theorem 21

$$\alpha + \beta \geq \beta.$$

Let γ be the smallest ordinal such that

(1) $$\alpha + \gamma \geq \beta.$$

I say that in fact

$$\alpha + \gamma = \beta,$$

for suppose that

$$\alpha + \gamma > \beta.$$

Clearly $\gamma \neq 0$, so there are two cases to consider. First, suppose γ has an immediate predecessor $\gamma - 1$. Then

$$\alpha + \gamma = \alpha + (\gamma - 1)^{\mathsf{I}} = (\alpha + (\gamma - 1))^{\mathsf{I}},$$

whence by Theorem 15 of §5.1

$$\alpha + \gamma > \alpha + (\gamma - 1) \geq \beta,$$

contrary to the assumption that γ is the smallest ordinal satisfying (1).

Second, suppose that γ is a limit ordinal. Then by the basic hypothesis on γ, for every $\delta < \gamma$

$$\alpha + \delta < \beta,$$

and hence by Theorem 63 of §2.6

$$\bigcup_{\delta \in \gamma} (\alpha + \delta) \subseteq \beta,$$

thus

$$\bigcup_{\delta \in \gamma} (\alpha + \delta) \leq \beta.$$

But

$$\alpha + \gamma = \bigcup_{\delta \in \gamma} (\alpha + \delta),$$

and so we have

$$\alpha + \gamma \leq \beta,$$

which combines with (1) to yield

$$\alpha + \gamma = \beta.$$

That γ is unique follows at once from the left cancellation law; that is, given

$$\alpha + \gamma_1 = \alpha + \gamma_2 = \beta,$$

by virtue of Theorem 19

$$\gamma_1 = \gamma_2. \qquad\qquad \text{Q.E.D.}$$

By the same method of argument used in the proof just completed, it is easy to establish:

THEOREM 24. *If $\beta \neq 0$ there is no ordinal γ such that for all $\delta < \beta$*

$$\alpha + \delta < \gamma < \alpha + \beta.$$

This theorem says, in other words, that $\alpha + \beta$ is the first ordinal greater than $\alpha + \delta$ for all $\delta < \beta$.

That ordinal addition is not commutative is shown by the simple counterexample:

$$1 + \omega \neq \omega + 1.$$

On the other hand, it is associative.

THEOREM 25. $\alpha + (\beta + \gamma) = (\alpha + \beta) + \gamma.$

PROOF. The proof proceeds by transfinite induction on γ.

Part 1. By virtue of Theorem 14

$$\alpha + (\beta + 0) = \alpha + \beta$$
$$= (\alpha + \beta) + 0.$$

Part 2. Our inductive hypothesis is that

$$(1) \qquad\qquad \alpha + (\beta + \gamma) = (\alpha + \beta) + \gamma,$$

and we then have the following identities:

$$\alpha + (\beta + \gamma') = \alpha + (\beta + \gamma)' \qquad \text{by Theorem 15}$$
$$= (\alpha + (\beta + \gamma))' \qquad \text{by Theorem 15}$$
$$= ((\alpha + \beta) + \gamma)' \qquad \text{by (1)}$$
$$= (\alpha + \beta) + \gamma' \qquad \text{by Theorem 15.}$$

Part 3. The inductive hypothesis is that γ is a limit ordinal and for every $\delta \in \gamma$

$$\alpha + (\beta + \delta) = (\alpha + \beta) + \delta.$$

From the definition of addition and the fact that $\beta + \gamma$ is a limit ordinal (Theorem 22), we have

$$\alpha + (\beta + \gamma) = \bigcup_{\theta \in \beta + \gamma} (\alpha + \theta)$$

$$= \bigcup \{\eta \colon (\exists \theta)(\theta \in \beta + \gamma \,\&\, \eta = \alpha + \theta\} \qquad \text{by definition.}$$

But now since each ordinal η has as members all smaller ordinals, we may replace in the last expression '$\theta \in \beta + \gamma$' by '$\beta + \delta \in \beta + \gamma$', and we obtain

$$\alpha + (\beta + \gamma) = \bigcup \{\eta \colon (\exists \delta)(\beta + \delta \in \beta + \gamma \,\&\, \eta = \alpha + (\beta + \delta))\}$$

$$= \bigcup \{\eta \colon (\exists \delta)(\delta \in \gamma \,\&\, \eta = \alpha + (\beta + \delta))\} \; \begin{array}{l}\text{by Theorem 17}\\\text{and Theorem 20}\end{array}$$

$$= \bigcup_{\delta \in \gamma} \alpha + (\beta + \delta) \qquad\qquad \text{by definition}$$

$$= \bigcup_{\delta \in \gamma} (\alpha + \beta) + \delta \qquad\qquad \begin{array}{l}\text{by inductive}\\\text{hypothesis}\end{array}$$

$$= (\alpha + \beta) + \gamma. \qquad\qquad\qquad \text{Q.E.D.}$$

We now turn to the definition of ordinal multiplication and a number of elementary theorems about this operation. The definition, like that for addition, simply generalizes the corresponding definition for integers. Note that the customary convention of letting juxtaposition denote multiplication is followed. Occasionally a dot is used for clarity.

DEFINITION 3. *The operation of ordinal multiplication is defined by the following recursion scheme*:

(i) $\alpha \cdot 0 = 0$,
(ii) $\alpha \cdot \beta^{\mathsf{I}} = \alpha \cdot \beta + \alpha$,
(iii) *if β is a limit ordinal*

$$\alpha \cdot \beta = \bigcup_{\gamma \in \beta} \alpha \cdot \gamma.$$

Proofs of several of the theorems are left as exercises.

THEOREM 26. $\alpha \cdot \beta$ *is an ordinal.*

THEOREM 27. $0 \cdot \alpha = 0$.

THEOREM 28. $\alpha \cdot 1 = 1 \cdot \alpha = \alpha$.

We now show that ordinal multiplication, like ordinal addition, is right-monotonic with respect to *less than*. In the case of multiplication we do have to add the condition that $\alpha > 0$.

THEOREM 29.   *If $\alpha > 0$ and $\beta < \gamma$ then $\alpha\beta < \alpha\gamma$.*

PROOF.   The proof, like that of the corresponding theorem for addition (Theorem 17), is by transfinite induction on γ.

Part 1.   If $\gamma = 0$ the hypothesis of the theorem is always false and thus the theorem holds vacuously.

Part 2.   The inductive hypothesis for this part is identical in appearance to the statement of the theorem. To prove the theorem holds for the successor of γ there are two cases to consider.

Case 1. $\beta < \gamma$.   Then we have:

$$\alpha\beta < \alpha\gamma,$$

but

$$\alpha\gamma' = \alpha\gamma + \alpha,$$

and since $\alpha > 0$ by virtue of Theorem 18

$$\alpha\gamma < \alpha\gamma',$$

whence by transitivity

$$\alpha\beta < \alpha\gamma'.$$

Case 2. $\beta = \gamma$.   Whence $\alpha\beta = \alpha\gamma$.   Then

$$\alpha\gamma' = \alpha\gamma + \alpha > \alpha\gamma.$$

Hence

$$\alpha\gamma' > \alpha\beta.$$

Part 3.   As in the case of the corresponding theorem for addition, the only part of the inductive hypothesis we need is that γ is a limit ordinal. Because $\beta < \gamma$ we have that $\beta \in \gamma$. But it is a fundamental property of ordinal multiplication that

$$\alpha\gamma = \bigcup_{\delta \in \gamma} \alpha\delta,$$

whence, by an argument exactly like that used in the proof of Theorem 17,

$$\alpha\beta \in \alpha\gamma,$$

i.e.,

$$\alpha\beta < \alpha\gamma. \qquad \text{Q.E.D.}$$

As an easy consequence of Theorem 29, we have:

THEOREM 30.   *If $\alpha \neq 0$ and $\beta \neq 0$ then $\alpha\beta \neq 0$.*

And as another immediate result we have a left cancellation law, provided $\alpha > 0$.

THEOREM 31.   *If $\alpha\beta = \alpha\gamma$ and $\alpha > 0$ then $\beta = \gamma$.*

A counterexample to right cancellation is given by the equality:

$$2\omega = 3\omega.$$

As an expected property of limit ordinals we have:

THEOREM 32.   *If β is a limit ordinal and $\alpha > 0$ then $\alpha\beta$ is a limit ordinal.*

The proof is left as an exercise, as is the corresponding "left dual."

THEOREM 33.   *If α is a limit ordinal and $\beta > 0$ then $\alpha\beta$ is a limit ordinal.*

We now prove the important fact that multiplication is distributive from the left with respect to addition.

THEOREM 34. $\alpha(\beta + \gamma) = \alpha\beta + \alpha\gamma.$

PROOF.   The proof proceeds by transfinite induction on γ. We assume $\alpha \neq 0$ (for Part 3), for otherwise the proof is trivial in view of Theorem 27.

Part 1. $\alpha(\beta + 0) = \alpha\beta = \alpha\beta + 0 = \alpha\beta + \alpha \cdot 0.$

Part 2.   We have the following identities:

$$
\begin{aligned}
\alpha(\beta + \gamma') &= \alpha(\beta + \gamma)' && \text{by Theorem 15} \\
&= \alpha(\beta + \gamma) + \alpha && \text{by (ii) of Definition 3} \\
&= (\alpha\beta + \alpha\gamma) + \alpha && \text{by inductive hypothesis} \\
&= \alpha\beta + (\alpha\gamma + \alpha) && \text{by associativity} \\
&= \alpha\beta + \alpha\gamma' && \text{by (ii) of Definition 3.}
\end{aligned}
$$

Part 3.   Suppose γ is a limit ordinal. Then by Theorem 22 $\beta + \gamma$ is also a limit ordinal, whence from the definition of multiplication

$$
\begin{aligned}
\alpha(\beta + \gamma) &= \bigcup_{\theta \in \beta + \gamma} \alpha\theta \\
&= \bigcup_{\substack{\theta \in \bigcup_{\delta \in \gamma}(\beta + \delta)}} \alpha\theta && \begin{array}{l}\text{by definition}\\\text{of addition}\end{array} \\
&= \bigcup \{\eta : (\exists\theta)(\theta \in \bigcup_{\delta \in \gamma}(\beta + \delta) \ \& \ \eta = \alpha\theta)\} && \text{by definition} \\
&= \bigcup \{\eta : (\exists\theta)(\exists\delta)(\delta \in \gamma \ \& \ \theta \in \beta + \delta \ \& \ \eta = \alpha\theta)\}.
\end{aligned}
$$

By virtue of Theorem 29, since if $\theta < \beta$ then $\alpha\theta < \alpha(\beta + \delta)$ for any $\delta \in \gamma$, and thus we may consider only θ's for which there is a $\delta \in \gamma$ such that $\theta = \beta + \delta$, because if $\theta_1 < \beta$ and $\theta_2 = \beta + \delta$ for some δ, then any member of $\alpha\theta_1$ is also a member of $\alpha\theta_2$ and the union of the indicated family of sets is unchanged by omitting θ_1. Whence we may replace the last identity above by:

$$\alpha(\beta + \gamma) = \bigcup \{\eta: (\exists\delta)(\delta \in \gamma \ \& \ \eta = \alpha(\beta + \delta)\}$$

$$= \bigcup_{\delta \in \gamma} \alpha(\beta + \delta) \qquad \text{by definition}$$

$$= \bigcup_{\delta \in \gamma} (\alpha\beta + \alpha\delta) \qquad \text{by inductive hypothesis}$$

$$= \bigcup \{\eta: (\exists\delta)(\delta \in \gamma \ \& \ \eta = \alpha\beta + \alpha\delta)\} \quad \text{by definition}$$

$$= \bigcup \{\eta: (\exists\theta)(\theta \in \alpha\gamma \ \& \ \eta = \alpha\beta + \theta)\} \quad \begin{array}{l}\text{by quantifier logic,}\\ \text{hypothesis that } \alpha \neq 0,\\ \text{and Theorem 29}\end{array}$$

$$= \bigcup_{\theta \in \alpha\gamma} (\alpha\beta + \theta) \qquad \text{by definition}$$

$$= \alpha\beta + \alpha\gamma \qquad \begin{array}{l}\text{by definition of}\\ \text{addition and Theorem}\\ \text{32.} \qquad\qquad \text{Q.E.D.}\end{array}$$

A simple example, which shows that multiplication is not distributive from the right, is the following, whose verification is left as an exercise:

$$(1 + 1)\omega \neq 1 \cdot \omega + 1 \cdot \omega.$$

The proof of the associativity of ordinal multiplication is by transfinite induction on γ; it is left as an exercise.

THEOREM 35. $\alpha(\beta\gamma) = (\alpha\beta)\gamma$.

From the character of the definitions of ordinal addition and multiplication, the definition of ordinal exponentiation is obvious.

DEFINITION 4. *The operation of ordinal exponentiation is defined by the following recursion scheme:*

(i) $\alpha^0 = 1$,

(ii) $\alpha^{\beta'} = \alpha^\beta \cdot \alpha$,

(iii) *if β is a limit ordinal and $\alpha > 0$ then*

$$\alpha^\beta = \bigcup_{\gamma \in \beta} \alpha^\gamma.$$

(iv) *if β is a limit ordinal and $\alpha = 0$ then*

$$\alpha^\beta = 0.$$

Proofs of the theorems on exponentiation are left as exercises.

THEOREM 36. $\alpha^1 = \alpha$.

THEOREM 37. *If $\beta > 0$ then $0^\beta = 0$.*

THEOREM 38. $1^\alpha = 1$.

THEOREM 39. $\alpha^{\beta+\gamma} = \alpha^\beta \cdot \alpha^\gamma$.

THEOREM 40. $(\alpha^\beta)^\gamma = \alpha^{\beta\gamma}$.

On the other hand, we do not always have:

$$(\alpha\beta)^\gamma = \alpha^\gamma \cdot \beta^\gamma,$$

for

$$(\omega \cdot 2)^2 \neq \omega^2 \cdot 2^2.$$

THEOREM 41. *If $\alpha > 1$ and $\beta < \gamma$ then $\alpha^\beta < \alpha^\gamma$.*

By transfinite induction on β we may also prove:

THEOREM 42. *If $\alpha > 1$ and $\beta > 1$ then*

$$\alpha + \beta \leq \alpha\beta \leq \alpha^\beta.$$

From the theorems and counterexamples given thus far in this section, it may be seen that the arithmetic of ordinal addition, multiplication, and exponentiation differs in three major respects from the arithmetic of the corresponding integer operations. (i) Ordinal addition and multiplication are not commutative. (ii) Ordinal multiplication is not distributive from the right with respect to ordinal addition, that is, we do not always have:

(1) $$(\alpha + \beta)\gamma = \alpha\gamma + \beta\gamma.$$

(iii) The exponent rule

(2) $$(\alpha \cdot \beta)^\gamma = \alpha^\gamma \beta^\gamma$$

sometimes fails. Of the three (i) is central, for if we assume that addition and multiplication are commutative we may derive (1) and (2). It is also worth mentioning that certain operations on 0 and 1 which are ordinarily undefined in analysis are definite in ordinal arithmetic. For example, 0^0, ∞^0, 1^∞ are not defined in analysis, but here:

$$0^0 = \omega^0 = \alpha^0 = 1,$$

$$1^\omega = 1^\beta = 1.$$

A number of investigations have been devoted to pursuing the development of ordinal arithmetic along the lines of classical results in number theory. Some material in this direction is given in the exercises at the end of the section. Here we restrict ourselves to stating that the analogues of Fermat's "Last Theorem" and Goldbach's Hypothesis are known to be false in ordinal number theory (Sierpinski [1950]). In elementary number theory Fermat's "Last Theorem" is the assertion that for $n \geq 3$ there are no natural numbers a, b, c such that

$$a^n + b^n = c^n.$$

The truth or falsity of this assertion is one of the famous open problems of number theory. The analogue for ordinal number theory is false. For any ordinal $\mu \geq 1$ there are ordinal numbers α, β, γ such that

$$\alpha^\mu + \beta^\mu = \gamma^\mu.$$

In particular, if μ has an immediate predecessor (i.e., μ is not a limit ordinal) then for $\xi \geq 1$

$$(\omega^\xi)^\mu + (\omega^\xi \cdot 2)^\mu = (\omega^\xi \cdot 3)^\mu.$$

If μ is a limit ordinal, then for $\xi \geq 1$

$$(\omega^\xi)^\mu + (\omega^{\xi \cdot \mu})^\mu = (\omega^{\xi \cdot \mu} + 1)^\mu.$$

Goldbach's Hypothesis is that every even natural number > 2 is the sum of two prime numbers. On the basis of the obvious definition of prime ordinal numbers, the hypothesis is false for ordinal numbers. It can be shown that $\omega + 10$ is not such a sum.

We now turn to the infinite sum and product operations. From a formal standpoint these two operations are unary operations on sequences of ordinals. We used the notion of μ-sequence in the proof of Theorem 7; we define it again here.

DEFINITION 5. *A μ-sequence of ordinals is a function whose domain is the ordinal μ and whose range is a set of ordinals.*

Sequences in the ordinary sense of analysis are ω-sequences. We use a standard notation for μ-sequences, namely, $\{\alpha_\xi\}_{\xi < \mu}$, where α_ξ is the ξ-th term of the sequence.

DEFINITION 6. *The infinite sum operation for ordinals is defined for any μ-sequence of ordinals $\{\alpha_\xi\}_{\xi < \mu}$ by the following recursion scheme:*

(i) $\displaystyle\sum_{\xi < 0} \alpha_\xi = 0,$

(ii) *if $\nu < \mu$ then*

$$\sum_{\xi < \nu'} \alpha_\xi = \sum_{\xi < \nu} \alpha_\xi + \alpha_\nu.$$

(iii) *if ν is a limit ordinal and $\nu \leq \mu$ then*

$$\sum_{\xi < \nu} \alpha_\xi = \bigcup_{\eta \in \nu} \sum_{\xi < \eta} \alpha_\xi.$$

It is perhaps worth noting the form the recursion scheme would take if we used a standard function notation for μ-sequences. Thus, let f be any sequence of ordinals. Then

$$\Sigma f \mid 0 = 0,$$

$$\Sigma f \mid \nu^{\mathsf{I}} = \Sigma f \mid \nu + f(\nu),$$

for ν a limit ordinal

$$\Sigma f \mid \nu = \bigcup_{\eta \in \nu} \Sigma f \mid \eta.$$

In spite of the greater simplicity of the latter notation, we shall in the sequel use the more customary notation introduced in Definition 6.

By transfinite induction we may prove that any ordinal may be obtained by iterated addition of 1.

THEOREM 43. *Let $\{\alpha_\xi\}_{\xi < \mu}$ be the μ-sequence such that for every $\xi < \mu$, $\alpha_\xi = 1$. Then*

$$\mu = \sum_{\xi < \mu} \alpha_\xi.$$

PROOF. Using Theorem 4, we proceed by induction on μ.

Part 1. If $\mu = 0$, then by (i) of the preceding definition

$$\sum_{\xi < 0} \alpha_\xi = 0.$$

Part 2. Suppose the theorem holds for μ. Now by (ii) of the definition

$$\sum_{\xi < \mu^{\mathsf{I}}} \alpha_\xi = \sum_{\xi < \mu} \alpha_\xi + 1$$

$$= \mu + 1 \qquad\qquad\qquad \text{by inductive hypothesis}$$

$$= \mu^{\mathsf{I}}.$$

Part 3. If μ is a limit ordinal, then by (iii) of the definition

$$\sum_{\xi < \mu} \alpha_\xi = \bigcup_{\eta \in \mu} \sum_{\xi < \eta} \alpha_\xi$$

$$= \bigcup_{\eta \in \mu} \eta \qquad\qquad\qquad \text{by inductive hypothesis}$$

$$= \mu \qquad\qquad\qquad \text{by Theorem 3. Q.E.D.}$$

We may also show that multiplication is iterated addition.

THEOREM 44. *Let $\{\alpha_\xi\}_{\xi < \beta}$ be the β-sequence such that for every $\xi < \beta$, $\alpha_\xi = \alpha$. Then*

$$\alpha \cdot \beta = \sum_{\xi < \beta} \alpha_\xi.$$

PROOF.   As in the last proof, we proceed by transfinite induction.

Part 1. $\beta = 0$.   By (i) of Definition 6

(1) $$\sum_{\xi < 0} \alpha_\xi = 0,$$

and by Theorem 27

(2) $$\alpha \cdot 0 = 0,$$

and thus (1) and (2) establish Part 1.

Part 2.   Suppose the theorem holds for β.   Then by (ii) of Definition 6

$$\sum_{\xi < \beta^!} \alpha_\xi = \sum_{\xi < \beta} \alpha_\xi + \alpha_\beta$$

$$= \sum_{\xi < \beta} \alpha_\xi + \alpha \qquad \text{by hypothesis}$$
of the theorem

$$= \alpha \cdot \beta + \alpha \qquad \text{by inductive hypothesis}$$

$$= \alpha \cdot \beta^! \qquad \text{by Definition 3.}$$

Part 3.   If β is a limit ordinal, then

$$\alpha \cdot \beta = \bigcup_{\gamma \in \beta} \alpha \cdot \gamma \qquad \text{by definition}$$
of multiplication

$$= \bigcup_{\gamma \in \beta} \sum_{\xi < \gamma} \alpha_\xi \qquad \text{by inductive hypothesis}$$

$$= \sum_{\xi < \beta} \alpha_\xi \qquad \text{by (iii) of}$$
Definition 6.   Q.E.D.

The following theorem is similar to Theorem 22 in its formulation and proof; so the proof is left as an exercise.

THEOREM 45.   *If μ is a limit ordinal and for every $\xi < \mu$, $\alpha_\xi \neq 0$, then $\sum_{\xi < \mu} \alpha_\xi$ is a limit ordinal.*

To formulate an associative law for the sum of a sequence of ordinals, there is a certain difficulty in finding a suitable notation.   To illustrate the ideas, consider the finite sum:

$$\alpha_0 + \alpha_1 + \alpha_2 + \alpha_3 + \alpha_4 + \alpha_5.$$

If

$$\gamma_0 = \alpha_0 + \alpha_1 + \alpha_2$$
$$\gamma_1 = \alpha_3 + \alpha_4$$
$$\gamma_2 = \alpha_5,$$

the general associative law asserts that

$$\gamma_0 + \gamma_1 + \gamma_2 = \alpha_0 + \alpha_1 + \alpha_2 + \alpha_3 + \alpha_4 + \alpha_5.$$

It is simpler to introduce a notation for the partial sums of the subscripts than for the γ_ξ directly. Let β_0 be the number of α_ξ terms of γ_0, that is, three, and in general β_ν be the number of α_ξ terms of γ_ν. Here

$$\beta_0 = 3$$
$$\beta_1 = 2$$
$$\beta_2 = 1.$$

Let σ_ν's be the partial sums of the β's. Thus

$$\sigma_\nu = \sum_{\eta < \nu} \beta_\eta.$$

In our example

$$\sigma_0 = 0$$
$$\sigma_1 = 3$$
$$\sigma_2 = \beta_0 + \beta_1 = 5$$
$$\sigma_3 = \beta_0 + \beta_1 + \beta_2 = 6.$$

Our example may then be represented by:

$$\sum_{\xi < 6} \alpha_\xi = \sum_{\nu < 3} \left(\sum_{\xi < \beta_\nu} \alpha_{\sigma_\nu + \xi} \right).$$

In this notation, we have the following general associative law.

THEOREM 46. *If* $\lambda = \sum_{\eta < \mu} \beta_\eta$ *and for* $\nu < \mu$, $\sigma_\nu = \sum_{\eta < \nu} \beta_\eta$, *then*

$$\sum_{\xi < \lambda} \alpha_\xi = \sum_{\nu < \mu} \left(\sum_{\xi < \beta_\nu} \alpha_{\sigma_\nu + \xi} \right).$$

PROOF. Without loss of generality we may assume that for every $\eta < \mu$, $\beta_\eta \neq 0$. We proceed by transfinite induction on μ.

Part 1. $\mu = 0$. The proof is immediate.

Part 2. Suppose the theorem holds for μ.

Then

$$\lambda = \sum_{\eta < \mu'} \beta_\eta = \sum_{\eta < \mu} \beta_\eta + \beta_\mu.$$

Moreover,

$$\sum_{\xi < \lambda} \alpha_\xi = \sum_{\xi < \sigma_\mu + \beta_\mu} \alpha_\xi$$

$$= \sum_{\xi < \sigma_\mu} \alpha_\xi + \sum_{\xi < \beta_\mu} \alpha_{\sigma_\mu + \xi}$$

$$= \sum_{\nu < \mu} \left(\sum_{\xi < \beta_\nu} \alpha_{\sigma_\nu + \xi} \right) + \sum_{\xi < \beta_\mu} \alpha_{\sigma_\mu + \xi} \qquad \text{by inductive hypothesis}$$

$$= \sum_{\nu < \mu'} \left(\sum_{\xi < \beta_\nu} \alpha_{\sigma_\nu + \xi} \right).$$

Part 3. Let μ be a limit ordinal. Then

(1)
$$\lambda = \sum_{\eta < \mu} \beta_\eta = \bigcup_{\delta < \mu} \sum_{\eta < \delta} \beta_\eta.$$

By virtue of Theorem 45, λ is a limit ordinal.

We then have:

$$\sum_{\xi < \lambda} \alpha_\xi = \bigcup_{\delta < \mu} \sum_{\xi < \sigma_\delta} \alpha_\xi \qquad \text{by (1) and (iii)}$$
of Definition 6

$$= \bigcup_{\delta < \mu} \sum_{\nu < \delta} (\sum_{\xi < \beta_\nu} \alpha_{\sigma_\nu + \xi}) \qquad \text{by inductive hypothesis}$$

$$= \sum_{\nu < \mu} (\sum_{\xi < \beta_\nu} \alpha_{\sigma_\nu + \xi}) \qquad \text{by (iii) of}$$
Definition 6.   Q.E.D.

The general associative law for addition may be used to prove a general distributive law, of which Theorem 34 is a special case. In this connection it should be noted that binary addition is obtained from the $\sum$ operation by considering only sequences of length two. The proof of this distributive law is left as an exercise.

THEOREM 47.
$$\alpha \cdot \sum_{\eta < \mu} \beta_\eta = \sum_{\eta < \mu} \alpha \cdot \beta_\eta.$$

Analogous to the definition of the sum of a sequence of ordinals, we define the *product* of such a sequence.

DEFINITION 7. *The infinite product operation for ordinals is defined for any μ-sequence of ordinals $\{\alpha_\xi\}_{\xi < \mu}$ by the following recursion scheme:*

(i) $\prod_{\xi < 0} \alpha_\xi = 1,$

(ii)   *if $\nu < \mu$ then*
$$\prod_{\xi < \nu'} \alpha_\xi = \prod_{\xi < \nu} \alpha_\xi \cdot \alpha_\nu,$$

(iii)   *if ν is a limit ordinal and $\nu \leq \mu$ then*
$$\prod_{\xi < \nu} \alpha_\xi = \bigcup_{\eta \in \nu} \prod_{\xi < \eta} \alpha_\xi,$$

unless $\alpha_\xi = 0$ for some $\xi < \nu$, in which case $\prod_{\xi < \nu} \alpha_\xi = 0$.

We state without proof some theorems on this infinite product operation. The proofs are analogous in structure to the ones given on infinite sums.

THEOREM 48. *Let $\{\alpha_\xi\}_{\xi < \beta}$ be the β-sequence such that for every $\xi < \beta$, $\alpha_\xi = \alpha$. Then*
$$\alpha^\beta = \prod_{\xi < \beta} \alpha_\xi.$$

The next theorem states the general associative law for products.

Theorem 49.   *If $\lambda = \sum\limits_{\eta < \mu} \beta_\eta$ and for $\nu < \mu$, $\sigma_\nu = \sum\limits_{\eta < \nu} \beta_\eta$, then*

$$\prod_{\xi < \lambda} \alpha_\xi = \prod_{\nu < \mu} \left(\prod_{\xi < \beta_\nu} \alpha_{\sigma_\nu + \xi} \right).$$

There is also a power rule, of which

$$\alpha^{\beta + \gamma} = \alpha^\beta \cdot \alpha^\gamma$$

is a special case.

Theorem 50.   *If $\lambda = \sum\limits_{\eta < \mu} \beta_\eta$ then $\alpha^\lambda = \prod\limits_{\eta < \mu} \alpha^{\beta_\eta}$.*

As the final theorem we summarize some interesting particular results, the proofs of which are not difficult.

Theorem 51.

(i)   *if for $\xi < \omega$, $\alpha_\xi = \xi + 1$, then*

$$\sum_{\xi < \omega} \alpha_\xi = 1 + 2 + 3 + \cdots = \omega$$

$$\prod_{\xi < \omega} \alpha_\xi = 1 \cdot 2 \cdot 3 \cdot \cdots = \omega,$$

(ii)   *if for $\xi < \omega$, $\alpha_\xi = n$ then*

$$\sum_{\xi < \omega} \alpha_\xi = n + n + n + \cdots = \omega \quad for \ n > 0,$$

$$\prod_{\xi < \omega} \alpha_\xi = n \cdot n \cdot n \cdot \cdots = \omega \quad for \ n > 1,$$

(iii)   *if for $\xi < \omega$, $\alpha_\xi = \omega$ then*

$$\sum_{\xi < \omega} \alpha_\xi = \omega + \omega + \omega + \cdots = \omega^2$$

$$\prod_{\xi < \omega} \alpha_\xi = \omega \cdot \omega \cdot \omega \cdot \cdots = \omega^\omega.$$

It is worth noting that the five operations on ordinal numbers defined in this section have essentially identical final clauses in their definiens concerning limit ordinals. We may define the notion of a transfinite sequence of ordinals having a limit, and then define continuity of a function of ordinals. Using these ideas, the clause just mentioned may be replaced by the requirement that the function be continuous (in the appropriate variable if it is a function of several variables).

We briefly indicate the relevant definitions. Let $\{\alpha_\xi\}_{\xi < \mu}$ be a μ-sequence of ordinals with μ a limit ordinal. Then α is the *limit* of this sequence, in symbols,

$$\alpha = \lim_{\xi < \mu} \alpha_\xi$$

if and only if for every $\xi < \mu$, $\alpha_\xi \leq \alpha$ and for every $\beta < \alpha$ there is a $\nu < \mu$ such that if $\nu < \xi$ then $\beta < \alpha_\xi$.

Let f be a function on some ordinal α and such that its range is a set of ordinals. Then f is *continuous* if and only if for every limit number $\lambda < \alpha$

$$f(\lambda) = \lim_{\xi < \lambda} f(\xi).$$

Thus, without using an exact notation, we may assert that $\alpha + \beta$ is continuous in β, and similarly for $\alpha \cdot \beta$ and α^β. On the other hand, these three operations are not continuous in α, for

$$\lim_{\xi < \omega} (\xi + 1) = \omega \neq \omega + 1$$

$$\lim_{\xi < \omega} (\xi \cdot 2) = \omega \neq \omega \cdot 2$$

$$\lim_{\xi < \omega} \xi^2 = \omega \neq \omega^2.$$

EXERCISES

1. Various counterexamples have been given without proof throughout this section. Making use of any theorems, prove the following:

    (a) $1 + \omega = 2 + \omega$
    (b) $n + \omega = \omega$
    (c) $1 + \omega < \omega + 1$
    (d) if $n \neq 0$ then $n + \omega < \omega + n$
    (e) $2\omega = 3\omega$
    (f) if $n \neq 0$ then $n\omega = \omega$
    (g) $2\omega < \omega \cdot 2$
    (h) $\omega \cdot 2 = \omega + \omega$
    (i) $(1 + 1)\omega < 1 \cdot \omega + 1 \cdot \omega$

2. Prove Theorems 20 and 21.

3. Is the converse of Theorem 22 true? If so, prove it. If not, give a counterexample.

4. Prove Theorem 27.

5. Prove Theorem 28.

6. Prove Theorems 30 and 31.

7. Prove Theorems 32 and 33.

8. Prove Theorem 35.

9. Prove the following two cancellation laws for ordinal multiplication:

    (a) If $\alpha\beta > \alpha\gamma$ then $\beta > \gamma$.
    (b) If $\alpha\gamma > \beta\gamma$ then $\alpha > \beta$.

10. Prove that if $\gamma < \alpha\beta$ then there is a unique $\alpha_1 < \alpha$ and a unique $\beta_1 < \beta$ such that $\gamma = \alpha\beta_1 + \alpha_1$.

11. If $\alpha > 0$ then for any ordinal β there is a unique γ and a unique $\delta < \alpha$ such that

$$\beta = \alpha\gamma + \delta.$$

12. If α is a limit ordinal then there is a unique β such that $\alpha = \omega\beta$.

13. Prove Theorem 36.
14. Prove Theorem 37.
15. Prove Theorem 38.
16. Prove Theorem 39.
17. Prove Theorem 40.
18. Show that $(\omega \cdot 2)^2 \neq \omega^2 \cdot 2^2$.
19. Prove Theorem 41.
20. Prove Theorem 42.
21. Prove Theorem 45.
22. Prove Theorem 47.
23. Prove Theorem 48.
24. Prove Theorem 49.
25. Prove Theorem 50.
26. Prove Theorem 51.
27. Prove that if for $\xi < \omega$, $\alpha_\xi = \omega^2$ then
$$\sum_{\xi < \omega} \alpha_\xi = \omega^2 + \omega^2 + \omega^2 + \cdots = \omega^3.$$
28. Prove that if for $\xi < \omega + \omega$, $\alpha_\xi = \omega$, then
$$\prod_{\xi < \omega + \omega} \alpha_\xi = \omega \cdot \omega \cdot \omega \cdots \omega \cdot \omega \cdot \omega \cdots = \omega^{\omega + \omega}.$$
29. Prove that if for $\xi < \omega$, $\alpha_\xi = \omega^\xi$, then
$$\sum_{\xi < \omega} \alpha_\xi = 1 + \omega + \omega^2 + \omega^3 + \cdots = \omega^\omega.$$

30. We define: β is a residual of γ if and only if $\beta > 0$ and there is an α such that $\alpha + \beta = \gamma$.
Prove that if $\alpha > \beta$ and α and β are residuals of γ then β is a residual of α.

31. We define: α is additively indecomposable if and only if $\alpha > 0$ and α is not the sum of two ordinals less than it.
   (a) What integers if any are additively indecomposable?
   (b) Prove that ω is additively indecomposable.
   (c) What is the smallest ordinal larger than ω which is additively indecomposable?
   (d) Using the definition of Exercise 30, characterize the residuals of an additively indecomposable ordinal.
   (e) Prove that if β is additively indecomposable and $\alpha < \beta$ then $\alpha + \beta = \beta$.
   (f) Prove that if $\beta > 1$ is additively indecomposable and $\alpha > 0$ then $\alpha\beta$ is additively indecomposable.
   (g) Prove that for every α, ω^α is additively indecomposable.

§ 7.3 Cardinal Numbers Again and Alephs.
Without using the special axiom for cardinals or the axiom of choice, a certain amount of cardinal number theory can be done by defining cardinal numbers as initial ordinals. An initial ordinal is one which is not equipollent to any ordinal less than it. However, without the axiom of choice we cannot show that every set has a cardinal number and thus cannot develop a reasonable cardinal arithmetic.

On the other hand, we can define the *alephs*. In classical intuitive set theory an aleph is the cardinal number of an infinite well-ordered set. This will not be our definition, but will be forthcoming as a theorem. In other words, we shall define alephs purely in terms of ordinal numbers, and then prove (in the next section) that the cardinal number of any infinite well-ordered set is an aleph. It is true that $\aleph_0$ was defined in Definition 28 of §5.3 as the cardinal number of ω (this definition depended on the special axiom for cardinals), but the real character of the alephs as the sequence of transfinite cardinal numbers was in no way indicated by this earlier definition.

To avoid any confusion it is to be emphasized that the material in this section is completely independent of §4.3, §4.4, and §5.3, which depend on the special axiom for cardinals.

DEFINITION 8. *x is a cardinal number if and only if x is an ordinal and for every ordinal α if $x \approx \alpha$ then $x \leq \alpha$.*

We have as two simple theorems:

THEOREM 52. *Every natural number is a cardinal number.*

THEOREM 53. *ω is a cardinal number.*

We also have without the axiom of choice the law of trichotomy for cardinal numbers.

THEOREM 54. *If α and β are cardinal numbers, then exactly one of the following: $\alpha < \beta, \beta < \alpha, \alpha = \beta$.*

PROOF. Since α and β are ordinals, either $\alpha \subseteq \beta$ or $\beta \subseteq \alpha$. Thus by Theorem 16 of §4.1 either $\alpha \leq \beta$ or $\beta \leq \alpha$, from which the theorem follows easily.   Q.E.D.

The fact that Theorem 54 may be proved without the axiom of choice does not mean that the trichotomy in terms of equipollence for arbitrary sets may be proved without that axiom.

We define transfinite cardinals in a way that agrees with Definition 27 of §5.3. The latter definition asserts that a transfinite cardinal is one which is the cardinal of a Dedekind infinite set. The present definition requires the cardinal to be $\geq \omega$, which implies that ω is a subset of it. But according to Theorems 47 and 48 of §5.3 all and only those sets which have a denumerable subset like ω are Dedekind infinite.

DEFINITION 9. *x is a transfinite cardinal if and only if x is a cardinal number and $x \geq \omega$.*

One of our objectives is to show that there is no largest transfinite cardinal. To this end, we associate with each set the set of ordinals equipollent to or less pollent than it. This notion will be the device which permits us to avoid the axiom of choice; the approach originates with Hartogs [1915], who uses it in his proof that the law of trichotomy implies that every set is well-ordered. (We also use it for the corresponding proof in the next chapter.)

Definition 10.

$$\mathcal{K}(A) = \{\alpha\colon \alpha \preceq A\}.$$

It requires a rather subtle use of the axiom schema of replacement to show that $\mathcal{K}(A)$ is not empty. The axiom schema of separation cannot be used, for there is no appropriate set of ordinals. Moreover, the application of the axiom schema of replacement is not direct, because to a given subset B of A there may be many distinct ordinals equipollent to it; that is, we cannot simply use, for B in the power set of A,

$$\varphi(B, \alpha) \text{ if and only if } B \approx \alpha.$$

An indirect but successful route is to consider *well-ordered* subsets of A, as the following proof shows.

Theorem 55. $\alpha \in \mathcal{K}(A)$ *if and only if* $\alpha \preceq A$.

proof.   To begin with, it is clear that

(1) $\alpha \preceq A$ if and only if there are sets B and R such that

   (i) $B \subseteq A$,
   (ii) R well-orders B,
   (iii) $\langle \alpha, \mathcal{E}\alpha \rangle$ is similar to $\langle B, R \rangle$.

(Similarity was defined in §5.1.)   We now define:

(2) $\mathcal{W}(A) = \{\langle B, R \rangle\colon B \subseteq A \mathbin{\&} R \text{ well-orders } B \mathbin{\&} R \subseteq A \times A\}.$

By virtue of the axiom schema of separation and the power set axiom it follows that

(3) $x \in \mathcal{W}(A)$ if and only if there are sets B and R such that

   (i) $x = \langle B, R \rangle$,
   (ii) $B \subseteq A$,
   (iii) R well-orders B.

Proof of (3) follows from taking as the basic set for the axiom schema of separation

$$(\mathcal{P}A) \times \mathcal{P}(A \times A),$$

since

(4) $$B \in \mathcal{P}A$$

and

(5) $$R \in \mathcal{P}(A \times A).$$

For application of the axiom schema of replacement we now take for φ:

(6) $\varphi(x,\alpha)$ if and only if x is similar to $\langle \alpha, \mathcal{E}\alpha \rangle$.

Clearly, from fundamental properties of ordinals, two distinct ordinals cannot be similar under the membership relation, and thus if $\varphi(x,\alpha)$ and $\varphi(x,\beta)$ then $\alpha = \beta$. Therefore by the axiom schema of replacement there is a set C such that

(7) $\alpha \in C \leftrightarrow (\exists x)(x \in \mathcal{W}(A) \ \& \ x \text{ is similar to } \langle \alpha, \mathcal{E}\alpha \rangle).$

From (1), (3), and (7) it follows at once that

(8) $\alpha \in C$ if and only if $\alpha \preceq A$,

and the theorem follows from (8).   Q.E.D.

THEOREM 56. $\mathfrak{IC}(A)$ *is an ordinal.*

PROOF.   If $\alpha \in \mathfrak{IC}(A)$ and $\beta < \alpha$ then $\beta \in \mathfrak{IC}(A)$, for if $\alpha \in \mathfrak{IC}(A)$, then there is a subset B of A such that $\alpha \approx B$ under some function, say f, but then $f \,|\, \beta$ guarantees that $\beta \in \mathfrak{IC}(A)$. Therefore $\mathfrak{IC}(A)$ is complete, and since it is a set of ordinals, it is connected by the membership relation. Whence being complete and connected, it is an ordinal.   Q.E.D.

The next theorem asserts that not $\mathfrak{IC}(A) \preceq A$; it is to be noted that the axiom of choice is needed to prove the stronger result $A < \mathfrak{IC}(A)$, although this can be established for well-ordered sets, and thus *a fortiori* for ordinals, without the axiom of choice.

THEOREM 57.   *Not* $\mathfrak{IC}(A) \preceq A$.

PROOF.   Because $\mathfrak{IC}(A)$ is an ordinal, by virtue of Theorem 55,

(1) $$\mathfrak{IC}(A) \preceq A \text{ if and only if } \mathfrak{IC}(A) \in \mathfrak{IC}(A),$$

but no ordinal is a member of itself, and consequently the theorem follows at once from (1).   Q.E.D.

We now show that $\mathfrak{IC}(A)$ is the smallest ordinal with the property of not being $\preceq A$.

THEOREM 58.   *If not* $\alpha \preceq A$ *then* $\mathfrak{IC}(A) \preceq \alpha$.

PROOF.   Suppose that $\alpha < \mathfrak{IC}(A)$. Then because $\mathfrak{IC}(A)$ is an ordinal, $\alpha \in \mathfrak{IC}(A)$; but then by virtue of the definition of $\mathfrak{IC}(A)$, $\alpha \preceq A$, which contradicts the hypothesis of the theorem.   Q.E.D.

It follows easily from the theorem just proved that

THEOREM 59. $\mathcal{K}(A)$ *is a cardinal number.*

Moreover, the theorems we have proved lead directly to the philosophically interesting consequence that

THEOREM 60.   *There is no largest cardinal number.*

PROOF.   Suppose if possible there were a largest cardinal, say α. Then by Theorem 57

$$(1) \qquad\qquad \text{not } \mathcal{K}(\alpha) \leq \alpha,$$

and by Theorem 59, $\mathcal{K}(\alpha)$ is a cardinal number, whence by Theorem 54 and (1)

$$\alpha < \mathcal{K}(\alpha),$$

which is contrary to our supposition about α. Q.E.D.

In order to define the alephs it is useful to introduce first the notion of the least ordinal satisfying a formula. Earlier consideration of this notion would have been possible and indeed would have occasionally been convenient.

DEFINITION SCHEMA 11.

$$\mu_\alpha(\varphi(\alpha)) = \beta \leftrightarrow [\varphi(\beta) \& (\forall\gamma)(\varphi(\gamma) \to \beta \leq \gamma)] \vee [\neg(\exists\alpha)\varphi(\alpha) \& \beta = 0].$$

Thus

$$\mu_\alpha(\alpha > 2) = 3,$$

$$\mu_\beta(\beta > \alpha) = \alpha + 1.$$

It is easily proved that

THEOREM SCHEMA 61.   *If* $\varphi(\beta)$ *then* $\mu_\alpha(\varphi(\alpha)) \leq \beta$.

THEOREM SCHEMA 62.   *If* $(\exists\alpha)\varphi(\alpha)$ *then* $\varphi(\mu_\alpha(\varphi(\alpha)))$.

We now define:

DEFINITION 12.   *The aleph operation is defined by the following recursion scheme:*[*]

  (i) $\aleph_0 = \omega$,
  (ii) $\aleph_{\alpha'} = \mu_\beta(\beta > \aleph_\alpha)$,
  (iii) *if* α *is a limit ordinal*

$$\aleph_\alpha = \bigcup_{\beta \in \alpha} \aleph_\beta.$$

We first prove:

[*]To conform to traditional notation we write '$\aleph_\alpha$' rather than '$\aleph(\alpha)$'.

THEOREM 63. *There is a β such that for every γ if $\gamma \in \alpha$ then $\beta > \aleph_\gamma$.*

PROOF. We need consider only $\alpha > 0$. By virtue of the axiom schema of replacement there is a non-empty set A such that

$$A = \{\eta: (\exists\gamma)(\gamma \in \alpha \,\&\, \eta = \aleph_\gamma)\}.$$

Now from earlier results on ordinals, $\cup A$ is an ordinal and moreover, if $\eta \in A$ then

$$\eta \leq \cup A.$$

Thus if there were no $\beta > \cup A$ then $\mu_\xi(\xi \leq \cup A \,\&\, \xi \approx \cup A)$ would be the largest transfinite cardinal, contrary to Theorem 60. Q.E.D.

It follows from this theorem, Definition 12, and some earlier results on ordinals that

THEOREM 64. *If $\alpha > 0$ then*

$$\aleph_\alpha = \mu_\beta((\forall\gamma)(\gamma \in \alpha \rightarrow \beta > \aleph_\gamma)).$$

This theorem, together with (i) of the definition, can be used as the definition of the aleph operation.

We restrict ourselves to a few further results, stated without proof. From Definition 12 and Theorems 63 and 64 it immediately follows that

THEOREM 65. *$\aleph_\alpha$ is a transfinite cardinal.*

THEOREM 66. *If $\alpha < \beta$ then $\aleph_\alpha < \aleph_\beta$.*

THEOREM 67. *There is no transfinite cardinal β such that $\aleph_\alpha < \beta < \aleph_{\alpha+1}$.*

Of greater difficulty is the proof that

THEOREM 68. *Every transfinite cardinal is an aleph, i.e., for every transfinite cardinal α there is a β such that $\alpha = \aleph_\beta$.*

We return briefly to the alephs at the end of the next section. We mention here that any ordinal number whose cardinality is $\aleph_0$ is said to be an *ordinal number of the second class*. (The finite ordinals are the numbers of the first class.) Many mathematical results are known about ordinal numbers of the second class, but many difficult problems concerning them are unsolved. (Some references are Sierpinski [1928], Church and Kleene [1937], Church [1938], Denjoy [1946].)

EXERCISES

1. Prove Theorems 52 and 53.
2. Give a detailed proof of (1) in the proof of Theorem 55.
3. Prove Theorem 59.

4. Prove that if $A \subseteq B$ then $\mathfrak{K}(A) \leq \mathfrak{K}(B)$.

5. Is it true that if $A \subseteq B$ then $\mathfrak{K}(A) < \mathfrak{K}(B)$?

6. Is it true that if $A \subset B$ then $\mathfrak{K}(A) < \mathfrak{K}(B)$?

7. Is it true that

$$\mathfrak{K}(A \cup B) = \mathfrak{K}(A) + \mathfrak{K}(B)?$$

8. Prove Theorem 61.

9. Prove Theorem 62.

10. Prove Theorem 64.

11. Prove Theorem 66.

12. Prove Theorem 67.

13. Prove that if α is a limit ordinal then there is no transfinite cardinal β such that for every $\gamma \in \alpha$, $\aleph_\gamma < \beta < \aleph_\alpha$.

14. For every cardinal α there is an ordinal β such that $\alpha < \aleph_\beta$.

15. Prove Theorem 68.

§ 7.4 Well-Ordered Sets.

To begin with, we may formulate the principle of transfinite induction for well-ordered sets. The proof is similar to that of Theorem 1. It should be noted that Theorem 69 implies Theorem 1.

THEOREM 69. [Principle of Transfinite Induction: Fourth Formulation]. *If*

(i) *R well-orders A,*

(ii) $(\forall y)[y \in A \ \& \ (\forall x)(x \in A \ \& \ x R y \rightarrow x \in B) \rightarrow y \in B]$,
then $A \subseteq B$.

Statement (and proof) of a principle for well-ordered sets similar to Theorem 4 is left as an exercise.

We state but do not prove that if an ordered set satisfies the principle of transfinite induction then it is well-ordered. Roughly speaking, this theorem combined with Theorem 69 shows that in order to apply transfinite induction to an ordered set it is necessary and sufficient that the set be well-ordered.

THEOREM 70. *Let R be a strict simple ordering of A and let A have an R-first element. Moreover, let A and R be such that for any set B, if $(\forall y)[y \in A \ \& \ (\forall x)(x \in A \ \& \ x R y \rightarrow x \in B) \rightarrow y \in B]$ then $A \subseteq B$. Then R well-orders A.*

Using the notion of a segment, which was introduced in Definition 32 of Chapter 3, we may formulate transfinite recursion in a manner similar to Theorem 8.

THEOREM SCHEMA 71. [Transfinite Recursion: Sixth Formulation]. *Let τ be any term and let R well-order A. Then there is a unique function F on A such that for every element x of A*

$$F(x) = \tau(F \mid \mathcal{S}(A,R,x)).$$

We shall use this theorem in Chapter 8 in the proof that if every set can be well-ordered then Zorn's Lemma holds.

The notion of an *increasing function* is familiar in numerical contexts, it easily generalizes to ordered sets.

DEFINITION 13. *f is an increasing function on* $\mathfrak{A} = \langle A, R \rangle$ *if and only if*

(i) *f is a function on A,*
(ii) $\mathfrak{R}f \subseteq A$,
(iii) *if* $x, y \in A$ *and* $x R y$ *then* $f(x) R f(y)$.

We now prove a fundamental theorem for such functions.

THEOREM 72. *If R well-orders A and f is an increasing function on* $\langle A, R \rangle$, *then there is no element x in A such that* $f(x)Rx$.

PROOF. Suppose, by way of contradiction, that there were an element x in A such that

$$(1) \qquad\qquad f(x)Rx.$$

Let

$$B = \{x \colon f(x) \, Rx\}.$$

By virtue of (1) $B \neq 0$, whence it has an R-first element, say x_1. Thus we have

$$(2) \qquad\qquad f(x_1)Rx_1.$$

Let

$$(3) \qquad\qquad x_0 = f(x_1),$$

whence

$$x_0 R x_1,$$

and since f is increasing

$$(4) \qquad\qquad f(x_0) R \, f(x_1).$$

From (3) and (4) we then have

$$f(x_0)Rx_0,$$

but then $x_0 \in B$ and x_1 is not the R-first element of B, which is absurd. Q.E.D.

We now state some theorems relating similarity (Definition 1 of §5.1) and increasing functions.

THEOREM 73. *If R well-orders A and f is an increasing function on* $\langle A, R \rangle$ *then* $\langle A, R \rangle$ *is similar under f to* $\langle \mathfrak{R}f, R \rangle$.

PROOF. We need to verify (a) − (c) of Definition 1 of §5.1; (b) follows at once from the definition of increasing functions. To establish (a), that is, that f is 1–1, consider any two distinct elements $x,y \in A$. Since $x \neq y$, we must have either $x\,R\,y$ or $y\,R\,x$. For definiteness, let it be $x\,R\,y$. But then $f(x)R\,f(y)$, and since R is irreflexive, $f(x) \neq f(y)$. Concerning (c) we have already that if $x\,R\,y$ then $f(x)R\,f(y)$. We need the converse. Suppose $f(x)R\,f(y)$. Then $x \neq y$ by virtue of the irreflexivity of R. If $y\,R\,x$ then $f(y)R\,f(x)$, but this is impossible since R is asymmetric. Whence, since R is connected, $x\,R\,y$. Q.E.D.

We leave as an exercise proof of what is essentially the converse.

THEOREM 74. *If R well-orders A, B is a subset of A, and $\langle A, R \rangle$ is similar under f to $\langle B, R \rangle$ then f is an increasing function on $\langle A, R \rangle$.*

We may use Theorems 72 and 74 to prove:

THEOREM 75. *If R well-orders A and $\langle A, R \rangle$ is similar to itself under f, then f is the identity function $\mathfrak{I}A$.*

PROOF. We need to show that for every x in A

$$f(x) = x.$$

On the basis of Theorems 72 and 74 there is no element x in A such that

(1) $$f(x)Rx.$$

Suppose now there is an element x in A such that

(2) $$x\,R\,f(x).$$

In view of the hypothesis of the theorem the inverse function f^{-1} is such that $\langle A, R \rangle$ is similar to itself under f^{-1}. Thus in view of Theorem 74

$$f^{-1}(x)R\,f^{-1}(f(x)),$$

that is,

(3) $$f^{-1}(x)Rx,$$

but (3) is impossible according to Theorem 72. Thus our supposition is false, and neither (1) nor (2) holds for any x in A. Since R is connected in A, it must follow that $f(x) = x$. Q.E.D.

It follows from the theorem just proved that

THEOREM 76. *If R well-orders A and S well-orders B, then there is at most one function f such that $\langle A, R \rangle$ is similar under f to $\langle B, S \rangle$.*

In other words, two well-ordered sets can be similar in at most one way.
Related application of preceding theorems yields:

THEOREM 77. *No well-ordered set is similar to one of its segments; that is, if R well-orders A then for any x in A, $\langle A, R \rangle$ is not similar to $\langle S(A,R,x),R \rangle$.*

We also have:

THEOREM 78. *If R well-orders A then the set of all R-segments of A ordered by proper inclusion is similar to $\langle A, R \rangle$.*

(Proper inclusion restricted to subsets of A is easily made a set-theoretical relation, say $\subset | A$, corresponding to $\mathscr{I}A$, $\mathcal{E}A$, $< | A$.)

THEOREM 79. *If*

   (i)   *R well-orders A,*
   (ii) *S well-orders B,*
   (iii) *for each x in A there is a y in B such that $\langle S(A,R,x),R \rangle$ is similar to $\langle S(B,S,y),S \rangle$,*
   (iv) *for each y in B there is an x in A such that $\langle S(B,S,y),S \rangle$ is similar to $\langle S(A,R,x),R \rangle$,*
then $\mathfrak{A} = \langle A, R \rangle$ is similar to $\mathfrak{B} = \langle B, S \rangle$.

PROOF. On the basis of Theorem 77, it is clear that (iii) and (iv) imply that there is a 1–1 function f from A onto B such that $\langle S(A,R,x),R \rangle$ is similar to $\langle S(B,S,f(x)),S \rangle$. We want to show that $\langle A, R \rangle$ is similar under f to $\langle B, S \rangle$. Let $x_1,x_2 \in A$, with $x_1 R x_2$. Then

$$(1) \qquad\qquad S(A,R,x_1) \subset S(A,R,x_2),$$

whence by hypothesis and Theorem 77,

$$(2) \qquad\qquad S(B,S,f(x_1)) \subset S(B,S,f(x_2))$$

and thus from the definition of segments

$$(3) \qquad\qquad f(x_1)S f(x_2).$$

Furthermore, (3) implies (2), which implies (1). Thus for $x_1,x_2 \in A$, $x_1 R x_2$ if and only if $f(x_1)S f(x_2)$, which establishes the theorem. Q.E.D.

We are now in a position to prove the *fundamental theorem for well-ordered sets*, namely, that either two well-ordered sets are similar or one is similar to a segment of the other. (It should be remarked that this theorem is the analogue for well-ordered sets of the law of trichotomy for unordered sets.) The proof of this fundamental theorem does not depend on the axiom schema of replacement.

THEOREM 80. *Let R well-order A and S well-order B. Then either*

   (i)   *$\langle A, R \rangle$ is similar to $\langle B, S \rangle$,*
   (ii) *there is an x in A such that $\langle S(A,R,x),R \rangle$ is similar to $\langle B, S \rangle$, or*
   (iii) *there is a y in B such that $\langle A, R \rangle$ is similar to $\langle S(B,S,y), S \rangle$.*

PROOF. Suppose $\langle A, R \rangle$ is not similar to $\langle B, S \rangle$. Then on the basis of Theorem 79 either there is a segment $S(A,R,x)$ with x in A which is not similar to any segment $S(B,S,y)$ for y in B, or there is a segment $S(B,S,y)$ not similar to any segment $S(A,R,x)$. We want to show the first alternative implies (ii); the proof that the second implies (iii) is identical.

Let $S(A,R,x_0)$ be the smallest such segment not similar to any segment $S(B,S,y)$ for some y in B. (Well-ordering of the segments of $\langle A, R \rangle$ or $\langle B, S \rangle$ follows from Theorem 78; in this connection also see Exercise 6 at the end of this section.) We want to show that every segment $S(B,S,y)$ is similar to some segment $S(A,R,x)$. Suppose, by way of contradiction, that there is a segment $S(B,S,y)$ not similar to any segment $S(A,R,x)$. Let $S(B,S,y_0)$ be the smallest such segment. Then every segment $S(B,S,y)$ of the set $S(B,S,y_0)$ is similar to a segment $S(A,R,x)$, and in every case

$$(1) \qquad\qquad S(A,R,x) \subset S(A,R,x_0),$$

for otherwise $S(A,R,x_0)$ would be similar to some segment $S(B,S,y)$, contrary to our hypothesis about it. By a corresponding argument every segment $S(A,R,x)$ of the set $S(A,R,x_0)$ is similar to some segment of the set $S(B,S,y_0)$: whence we may use Theorem 79 to show that $S(A,R,x_0)$ is similar to $S(B,S,y_0)$. But this similarity contradicts the definition of $S(A,R,x_0)$, and we conclude that our supposition that there is a segment such as $S(B,S,y_0)$ is false. Consequently we conclude that every segment $S(B,S,y)$ is similar to some segment $S(A,R,x)$ and by the argument establishing (1), $S(A,R,x)$ must be a segment of $S(A,R,x_0)$. Since $S(A,R,x_0)$ is the smallest segment of A not similar to some segment of B, we conclude by virtue of Theorem 79 again that $\langle (S(A,R,x_0)),R \rangle$ is similar to $\langle B, S \rangle$, which establishes (ii), granted the first of the two alternatives stated at the beginning of the proof. Q.E.D.

We now prove a representation theorem for well-ordered sets in terms of ordinal numbers. The proof uses the axiom schema of replacement.

THEOREM 81. *If R well-orders A then there is a unique ordinal α such that $\langle A, R \rangle$ is similar to $\langle \alpha, \mathcal{E}\alpha \rangle$.*

PROOF. The idea of the proof stems from the informal discussion in §5.1 directed toward justifying intuitively the von Neumann construction of the ordinals. We use the sixth formulation of transfinite recursion (Theorem 71) to define on A a function f such that for x in A

$$(1) \qquad\qquad f(x) = \mathcal{R}(f|S(A,R,x)).$$

We must show that f is 1–1. Suppose not. Then let y be the least element under the ordering R of A such that for some x, $x\,R\,y$ and $f(x) = f(y)$. It follows at once from (1) that then

$$(2) \qquad\qquad f(x) = f(x'),$$

where x' is the immediate successor of x under the ordering R; but it is also clear from (1) that

(3) $$f(x') = f(x) \cup \{f(x)\},$$

but (2) and (3) are jointly absurd, whence we conclude:

(4) $$f \text{ is } 1\text{-}1.$$

Secondly, it is not difficult to show that $f(x)$ is an ordinal, namely, that it is complete and connected by the membership relation. Let B be an element of $f(x)$, i.e.

$$B \in \mathcal{R}(f|\mathcal{S}(A,R,x)),$$

whence there must be a y in $\mathcal{S}(A,R,x)$, i.e., $y \, R \, x$, such that $f(y) = B$, but clearly if $y \, R \, x$ then $\mathcal{S}(A,R,y) \subset \mathcal{S}(A,R,x)$ and thus

$$B = \mathcal{R}(f|\mathcal{S}(A,R,y)) \subseteq \mathcal{R}(f|\mathcal{S}(A,R,x)),$$

which shows that B is a subset of $f(x)$, and consequently $f(x)$ is complete. To show that $f(x)$ is connected by the membership relation, let $A,B \in f(x)$. Then there are elements a and b in $\mathcal{S}(A,R,x)$ such that $f(a) = A$ and $f(b) = B$. Suppose $A \neq B$. Then $a \neq b$ and by the connectivity of R in A, either aRb or bRa. If aRb then by virtue of (1) and the 1–1 character of f, $f(a) \in f(b)$. If bRa then $f(b) \in f(a)$, whence $f(x)$ is connected by the membership relation. Moreover, the argument just given shows that if $x \, R \, y$ then $f(x) \in f(y)$. On the other hand, if $f(x) \in f(y)$ then $x \, R \, y$. For suppose not. Then either $x = y$ or $y \, R \, x$, from the connectivity of R. If the former then $f(x) = f(y)$; and if the latter, $f(y) \in f(x)$; either conclusion is absurd on the hypothesis that $f(x) \in f(y)$. Thus we have established for any x and y in A

(5) $$x \, R \, y \text{ if and only if } f(x) \in f(y).$$

Finally, we want to show that the range of the function f, $\mathcal{R}(f)$, is an ordinal. Since every member of $\mathcal{R}(f)$ is an ordinal, $\mathcal{R}(f)$ is connected by the membership relation. Let α be a member of $\mathcal{R}(f)$. We prove by transfinite induction that α is a subset of $\mathcal{R}(f)$. It follows at once from (1) that $0 \in \mathcal{R}(f)$. Secondly, if $\beta < \alpha$ and $\beta \in \mathcal{R}(f)$, then there is an x in A such that $f(x) = \beta$, and $f(x') = \beta^{\mathfrak{l}}$ on the basis of (3), whence $\beta^{\mathfrak{l}} \in \mathcal{R}(f)$. Finally if β is a limit ordinal, $\beta \leq \alpha$ and for every $\gamma < \beta$, $\gamma \in \mathcal{R}(f)$, then

$$\bigcup_{\gamma < \beta} \mathcal{S}(A,R,f^{-1}(\gamma))$$

is an R-segment of A; and consequently there is a y in A such that $\beta = f(y)$, whence $\beta \in \mathcal{R}(f)$. We have thus established that

(6) $$\mathcal{R}(f) \text{ is an ordinal.}$$

Our theorem follows at once from (4), (5), and (6), for the uniqueness of the ordinal $\Re(f)$ is obvious. Q.E.D.

Note that the fundamental theorem for well-ordered sets (Theorem 80) follows rather directly from this representation theorem.

We conclude this section with a few theorems on cardinals and alephs. From the representation theorem just proved it easily follows that

THEOREM 82. *If R well-orders A then there is a unique cardinal number α such that $\alpha \approx A$.*

This theorem justifies the definition of the cardinal number of a well-ordered set.

DEFINITION 14. *If R well-orders A then $\overline{\overline{A}} = x$ if and only if x is a cardinal number and $x \approx A$.*

Some obvious theorems are:

THEOREM 83. *If R well-orders A and S well-orders B then $\overline{\overline{A}} = \overline{\overline{B}}$ if and only if $A \approx B$.*

THEOREM 84. $\overline{\overline{\alpha}} \leq \alpha$.

Finally, we may combine the results of Theorems 68, 81, and 82 to yield the classical result:

THEOREM 85. *The cardinal number of an infinite, well-ordered set is an aleph.*

EXERCISES

1. Prove Theorem 69.
2. Formulate and prove for well-ordered sets an analogue of Theorem 4.
3. Prove Theorem 70.
4. Prove Theorem 71.
5. Prove Theorem 74.
6. Let R well-order A and let $\langle A, R \rangle$ be similar to $\langle B, S \rangle$. Prove that S well-orders B.
7. Let R be a strict simple ordering of A, and let f be an increasing function on $\langle A, R \rangle$.
    (i) Is f a 1-1 function?
    (ii) If so, is the inverse of f an increasing function?
8. Prove Theorem 76.
9. Let R and S be strict simple orderings of A and let f be an increasing function on both $\langle A, R \rangle$ and $\langle A, S \rangle$. How are R and S related?
10. Prove Theorem 77.
11. Theorems 72–77 have the hypothesis that R well-orders A. In which of these theorems may this hypothesis be weakened to: R is a strict simple ordering of A?

12. Prove that if R well-orders A and $x,y \in A$ then
   (i) either $\mathcal{S}(A,R,x) \subseteq \mathcal{S}(A,R,y)$ or $\mathcal{S}(A,R,y) \subseteq \mathcal{S}(A,R,x)$,
   (ii) $x\, R\, y$ if and only if $\mathcal{S}(A,R,x) \subset \mathcal{S}(A,R,y)$,
   (iii) if $x \neq y$ then $\mathcal{S}(A,R,x)$ is not similar to $\mathcal{S}(A,R,y)$.

13. Prove Theorem 78.

14. Prove that if R well-orders A, S well-orders B, and $\langle A, R \rangle$ is similar to $\langle B, S \rangle$, then for each x in A there is a y in B such that $\langle \mathcal{S}(A,R,x),R \rangle$ is similar to $\langle \mathcal{S}(B,S,y),S \rangle$.

15. Give a counterexample to show that Theorem 79 does not hold if the hypothesis of well-ordering is weakened to strict simple ordering.

16. Let R well-order A and let B be a subset of A. Prove that $\langle B, R \rangle$ is similar to $\langle A, R \rangle$ or to $\langle \mathcal{S}(A,R,x),R \rangle$ for some x in A.

17. Define

$$M(\alpha) = \{x: x \text{ is a transfinite cardinal } \& \ x \in \alpha\}.$$

Prove that for every β there is a unique transfinite cardinal α such that $\langle \beta, \mathcal{E}\beta \rangle$ is similar to $\langle M(\alpha), \mathcal{E}\alpha \rangle$. (This result justifies an alternative definition of the alephs.)

§ 7.5 Revised Summary of Axioms.

In view of the fact that two of our earlier axioms may be derived from the axiom schema of replacement and the power set axiom, this seems an appropriate point at which to revise the summary of §2.10. We first show:

METATHEOREM 1. *The axiom schema of separation is derivable from the axiom schema of replacement.*

PROOF. In the axiom schema of replacement take $\varphi(x,y)$ to be $x = y$ & $\psi(y)$, which is clearly functional in x. We then have:

$$(\exists B)(\forall y)(y \in B \leftrightarrow (\exists x)(x \in A \ \& \ x = y \ \& \ \psi(y))),$$

from which the axiom schema of separation follows by elementary quantifier logic and the logic of identity. Q.E.D.

Given the axiom schema of separation it might seem that by the addition of a finite number of axioms (*not* axiom schemata) Zermelo set theory could be extended to Zermelo-Fraenkel set theory; but Montague [1956] has shown that no such finite extension is equivalent to adding the axiom schema of replacement.

The following result has been mentioned earlier; it is referred to in Zermelo [1930].

METATHEOREM 2. *The pairing axiom is derivable from the power set axiom and the axiom schema of replacement.*

PROOF. In the axiom schema of replacement we select as the set A, the power set

(1) $$\mathcal{P}\mathcal{P}(0) = \{0, \{0\}\}.$$

Let x,y be the two objects whose pair set we want to form. As $\varphi(u,v)$ in the axiom schema of replacement (we have changed the variables in φ to avoid confusion), we take:

(2) $\qquad\qquad (u = 0 \,\&\, v = x) \lor (u = \{0\} \,\&\, v = y).$

Clearly (2) is functional in u, that is, for each u in $\mathcal{P}\mathcal{P}(0)$ there is exactly one v such that $\varphi(u,v)$. Applying as indicated the axiom schema of replacement to (1) and (2), we just obtain that there is a set B with exactly x and y as members; that is, from $(\exists B)(\forall v)(v \in B \leftrightarrow (\exists u)(u \in \mathcal{P}\mathcal{P}(0) \,\&\, [(u = 0 \,\&\, v = x) \lor (u = \{0\} \,\&\, v = y)]))$ we infer the pairing axiom:

$$(\exists A)(\forall z)(z \in A \leftrightarrow z = x \lor z = y). \qquad \text{Q.E.D.}$$

With these two derivations at hand, the system of set theory developed in these first seven chapters depends on just seven axioms. Moreover since the axiom for cardinals is not a standard axiom of Zermelo set theory, every use of it is indicated by a dagger. The seven axioms in chronological order of introduction are:

AXIOM OF EXTENSIONALITY:

$$(\forall x)(x \in A \leftrightarrow x \in B) \to A = B.$$

SUM AXIOM:

$$(\exists C)(\forall x)(x \in C \leftrightarrow (\exists B)(x \in B \,\&\, B \in A)).$$

POWER SET AXIOM:

$$(\exists B)(\forall C)(C \in B \leftrightarrow C \subseteq A).$$

AXIOM OF REGULARITY:

$$A \neq 0 \to (\exists x)[x \in A \,\&\, (\forall y)(y \in x \to y \notin A)].$$

AXIOM FOR CARDINALS:

$$\mathcal{K}(A) = \mathcal{K}(B) \leftrightarrow A \approx B.$$

AXIOM OF INFINITY:

$$(\exists A)(0 \in A \,\&\, (\forall B)(B \in A \to B \cup \{B\} \in A)).$$

AXIOM SCHEMA OF REPLACEMENT: *If* $(\forall x)(\forall y)(\forall z)(x \in A \,\&\, \varphi(x,y)$ *&* $\varphi(x,z) \to y = z)$ *then* $(\exists B)(\forall y)(y \in B \leftrightarrow (\exists x)(x \in A \,\&\, \varphi(x,y))).$

CHAPTER 8

THE AXIOM OF CHOICE

§ 8.1 Some Applications of the Axiom of Choice. In Chapters 4
5, and 7 various results have been mentioned which require the axiom o
choice for their proof. The formulation of the axiom which we shall use is

*For any set A there is a function f such that for any non-empty subse
B of A, f(B) ∈ B.*

For future reference we label this formulation: AC_1. The function f i
often called a choice function or a selection function for the given set A
The function includes in its domain every non-empty subset of A an
it selects exactly one element from each such subset. To acquire som
feeling for this choice function we may consider a simple finite example
for which the axiom is not needed. Let

$$A = \{1,2\}$$
$$B_1 = \{1\}$$
$$B_2 = \{2\}.$$

Then there are two distinct choice functions f_1 and f_2 whose domains a
the non-empty subsets of A:

$$f_1(B_1) = f_2(B_1) = 1$$
$$f_1(B_2) = f_2(B_2) = 2$$
$$f_1(A) = 1$$
$$f_2(A) = 2.$$

This notion of a choice function is easily formalized.

DEFINITION 1. *f is a choice function for A if and only if f is a functio
whose domain is the family of non-empty subsets of A and for eve
$B \subseteq A$ with $B \neq 0$, $f(B) \in B$.*

Using this definition the axiom of choice may be·formulated. *Every set has a choice function*, although in the formulation given above of the axiom it was not required that the domain of *f* be restricted to non-empty subsets of *A*. However, this is a trivial difference.

Every theorem or definition which depends on the axiom of choice is indicated by ' ★ '. Without the axiom we may prove the following two theorems about choice functions. We leave the proofs, which are not difficult, as exercises.

THEOREM 1.   *If R well-orders A then A has a choice function.*

THEOREM 2.   *Every finite set has a choice function.*

Historically the axiom of choice was first introduced by Zermelo [1904] in order to prove that every set can be well-ordered. Until the last two or three decades probably the main application of the axiom in general mathematics was through the well-ordering theorem and the application of transfinite induction to the well-ordering guaranteed by the theorem. However, the recent trend among mathematicians has been to avoid transfinite induction and use some maximal principle. (Precise formulations are given in the next section.)

Specific applications of the axiom of choice have been mentioned in preceding chapters. In §4.1 it was mentioned that the axiom was needed to prove the law of trichotomy for the pollence of sets, that is, $A \prec B$, $B \prec A$, or $A \approx B$, for any two sets *A* and *B*. In §4.2 it was asserted that every known proof that ordinary infinity implies Dedekind infinity requires the axiom. In §4.3 the necessity was mentioned of using the axiom to prove the law of trichotomy for cardinal numbers constructed by use of the special axiom for them. Obviously the trichotomy for pollence of sets implies this result immediately. In §5.3 we noted that the axiom is needed to show that every infinite set has a denumerable subset. This fact is the crucial one needed in proving that ordinary infinity implies Dedekind infinity. Finally, it was stated in §7.3 that the proof that every set has a cardinal number requires the axiom, when cardinal numbers are defined as initial ordinals.

We turn now to the proofs of these facts, which shall not necessarily be given here in the order in which they occurred in the text.

We begin with

★THEOREM 3.★   *If a set is infinite then it has a denumerable subset.*

PROOF. Let *A* be an infinite set and let *f* be a choice function for *A* as postulated by the axiom of choice. We now use Theorem 27 of §5.2 to

---

*We note again that we star all theorems and definitions which depend on the axiom of choice.

define a unique function g on ω:

$$g(0) = f(A)$$

$$g(n^!) = f(A \sim \{g(k) \colon k \leq n\}).$$

(Thus, for instance,

$$g(1) = f(A \sim \{g(0)\}) = f(A \sim \{f(A)\}).)$$

The intuitive idea is that g assigns 0 to the element x_0 selected from A, then 1 to the element x_1 selected from $A \sim \{x_0\}$, then 2 to the element x_2 selected from $A \sim \{x_0, x_1\}$, etc. Since always $f(B) \in B$, we see that g is 1–1. Now suppose there is an n such that

$$A \sim \{g(k) \colon k < n\} = 0;$$

then g establishes that

$$A \approx n,$$

contrary to the hypothesis that A is infinite. Hence for every n

$$A \sim \{g(k) \colon k < n\} \neq 0,$$

whence we conclude that the range of g is equipollent with ω, but the range of g is a subset of A, which means that the theorem is established. Q.E.D.

From the theorem just proved and Theorems 46 and 47 of §5.3 we conclude:

⋆Theorem 4. *A set is Dedekind infinite if and only if it is infinite.*

For subsequent purposes, we now prove what Bernays ([1942], p. 141) appropriately calls a *numeration theorem*: every set can be placed in 1–1 correspondence with some ordinal.

⋆Theorem 5. [Numeration Theorem]. *For any set A there is an ordinal α such that $\alpha \approx A$.*

proof. By virtue of the axiom of choice there is a function F such that for every non-empty subset B of A

$$F(B) \in B.$$

We use F to define the following predicate φ: $\varphi(B, \beta)$ if and only if*

(1) $B \subseteq A$,

there is an f such that

(2) f is a function on β,

(3) $\Re f = B$,

(4) $(\forall \gamma)(\gamma < \beta \rightarrow f(\gamma) = F(A \sim \Re(f|\gamma))$.

---

*In this definition we assume that $0 \notin A$. The case of $0 \in A$ can be handled separately later.

Obviously f, if it exists, is unique and is 1–1, whence if $\varphi(B,\beta)$ and $\varphi(B,\beta_1)$ then $\beta = \beta_1$ and we may apply the axiom schema of replacement to form the set C such that

$$\beta \in C \leftrightarrow (\exists B)(B \in \mathcal{P}A \ \& \ \varphi(B,\beta)).$$

The remainder of the proof follows along lines familiar from §7.1 on transfinite induction by considering $\cup C$. The details are left as an exercise. Q.E.D.

By using the numeration theorem the following two theorems may be easily proved.

★THEOREM 6. *Every set can be well-ordered; that is, for every set A there is a relation R such that R well-orders A.*

★THEOREM 7. [Trichotomy.] *For any two sets A and B, $A \prec B$, $B \prec A$, or $A \approx B$.*

As another easy consequence of the numeration theorem, we have:

★THEOREM 8. *For any set A there is a unique cardinal number α such that $\alpha \approx A$.*

And this theorem justifies introducing Cantor's double-bar notation for the *cardinal number* of a set.

★DEFINITION 2. $\overline{\overline{A}} = x$ *if and only if x is a cardinal number and $x \approx A$.*

An obvious theorem is:

★THEOREM 9. $\overline{\overline{A}} = \overline{\overline{B}}$ *if and only if $A \approx B$.*

On the basis of Theorem 9, we may thus regard the special axiom for cardinals as redundant once we adopt the axiom of choice. That is to say, this special axiom is derivable from our other axioms together with the axiom of choice.

EXERCISES

1. How many distinct choice functions are there
   (a) for a set of three elements?
   (b) for a set of n elements?
2. What is the obvious choice function for the set of negative integers?
3. Prove Theorem 1.
4. Prove Theorem 2.
5. Complete the proof of Theorem 5.
6. Prove Theorem 6.

7.   Prove Theorem 7.

8.   Use the axiom of choice to prove that if A is a denumerable family of pairwise disjoint, finite sets then $\cup A$ is denumerable.

9.   Prove Theorem 8.

10.   Prove Theorem 9.

11.   Prove that if f is a function then $\overline{\overline{\Re f}} \leq \overline{\overline{\mathfrak{D} f}}$.

§ 8.2 Equivalents of the Axiom of Choice.

In his classical paper of 1904 Zermelo was concerned to show that the axiom of choice implies that every set can be well-ordered. In that paper he used the following formulation:

AC_2: *If A is a set of non-empty, pairwise disjoint sets, then there is a set C whose intersection with any member B of A has exactly one element, that is, $C \cap B$ is a unit set.*

Our objective in this section is to state a number of principles equivalent to AC_1; many of the proofs will be left as exercises. We begin with:

THEOREM 10. AC_2 *is equivalent to* AC_1.

Another common formulation is:

AC_3: *Given any relation R there is a function $f \subseteq R$ such that*

$$domain\ of\ R = domain\ of\ f.$$

THEOREM 11. AC_3 *is equivalent to* AC_1.

Our program is now to establish the following equivalences:

Axiom of choice ↔ numeration theorem
↔ well-ordering theorem
↔ Zorn's Lemma
↔ law of trichotomy.

As would be expected, none of the theorems establishing these equivalences depend on the axiom of choice, and consequently they are all unstarred. In the last section (Theorem 5) we proved:

Axiom of choice → numeration theorem.

Furthermore the obvious approach to proving Theorems 6 and 7 yields:

THEOREM 12.   *The numeration theorem implies the well-ordering theorem and the law of trichotomy.*

Our next step is to prove that the well-ordering theorem implies Zorn's Lemma. Some preliminary definitions are needed:

Definition 3.

(i)   *A is a chain if and only if A is a set of sets and for any two sets B and C in A either $B \subseteq C$ or $C \subseteq B$.*

(ii)   *A is a chain in B if and only if A is a chain and $A \subseteq B$.*

(iii) *A is a maximal chain in B if and only if A is a chain in B and there is no chain C in B such that $A \subset C$.*

The intuitive idea is just that a chain is a set which is simply ordered by inclusion.* As for maximal chains, consider the set $\mathcal{P}\omega$ of subsets of integers.

Then the family of sets

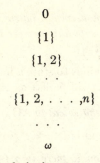

$$0$$
$$\{1\}$$
$$\{1, 2\}$$
$$\cdots$$
$$\{1, 2, \ldots ,n\}$$
$$\cdots$$
$$\omega$$

is an example of a maximal chain in $\mathcal{P}\omega$.

We also need to recall (Definition 4 of Chapter 4) the notion of a maximal element of a set A. For instance, if

$$A = \{\{1\}, \{1, 2\}, \{3\}\}$$

then $\{1, 2\}$ and $\{3\}$ are maximal elements of A. If we let

$$A = \mathcal{P}B$$

then B is the unique maximal element of A. If we let

$$A = (\mathcal{P}\omega) \sim \{\omega\},$$

then A has an infinity of maximal elements, namely the sets

$$\omega \sim \{1\}$$
$$\omega \sim \{2\}$$
$$\cdots$$
$$\cdots$$

The definition of maximal element is such as to require the element to be a set.

We may now state

*What is here called a *chain* is called a *nest* in Kelley [1955].

ZORN'S LEMMA Z_1: *If $A \neq 0$ and if the sum of each non-empty chain which is a subset of A is in A, then A has a maximal element.*

Formulated symbolically, Z_1 is:

$A \neq 0$ & $(\forall B)(B \subseteq A$ & B is a non-empty chain $\rightarrow \cup B \in A) \rightarrow A$ has a maximal element.

This maximal principle is baptized after Zorn [1935], but the history of it and some closely related maximal principles is very tangled. Certainly Zorn was essentially anticipated by F. Hausdorff, C. Kuratowski, and R. L. Moore at the least. Several variant formulations of Zorn's Lemma are given below, particularly the Maximal Principle of Hausdorff, which dates from 1914.

To continue our implications we want to prove:

THEOREM 13. *The well-ordering theorem implies Zorn's Lemma Z_1.*

PROOF. The intuitive idea of the proof is to use the postulated well-ordering R to build, for any non-empty family A of sets, a maximal chain: the R-first element of A is in this chain, and so is any subsequent element which either includes or is included in every R-preceding element already in the chain. The sum of this maximal chain is a maximal element as desired.

Let A be a set satisfying the hypothesis of Z_1, and let R be a well-ordering of A as postulated. We also make the simplifying, inessential assumption that A has only sets as members.

The first step in "constructing" a maximal chain is to define a term τ which may be used to define a function by transfinite recursion which picks out members of this maximal chain.

We define:

$$\tau(h) = \begin{cases} 1, & \text{if for every } B \text{ in } \mathfrak{D}h \text{ such that } h(B) = 1, \\ & \text{we have} \\ & \qquad B \subseteq D \text{ or } D \subseteq B, \\ & \text{where } D \text{ is the } R\text{-first element of } A \sim \mathfrak{D}h; \\ 0, & \text{otherwise.} \end{cases}$$

The second step is to apply transfinite recursion to τ and R. On the basis of the sixth formulation of transfinite recursion (Theorem 71 of §7.4) we know that there is a unique function f such that

(i) f is a function on A
(ii)   for every set $B \in A$

$$f(B) = \tau(f \mid \mathcal{S}(A,R,B)).$$

The third step is to define the appropriate chain. Let

$$C = \{B: B \in A \mathbin{\&} f(B) = 1\}.$$

To verify that C is a chain we need to show that for any two members B_1 and B_2 of C either

$$B_1 \subseteq B_2 \quad \text{or} \quad B_2 \subseteq B_1.$$

For definiteness let B_1 R-precede B_2 in the well-ordering of A. Since $B_1 \in C$,

(1) $$f(B_1) = 1,$$

also by hypothesis

$$f(B_2) = 1,$$

but then

$$\tau(f \mid \mathcal{S}(A,R,B_2)) = 1,$$

and B_2 is the R-first element of

$$A \sim \mathfrak{D}(f \mid \mathcal{S}(A,R,B_2)),$$

whence since $B_1 \in \mathcal{S}(A,R,B_2)$, by the definition of τ and (1)

$$B_1 \subseteq B_2 \quad \text{or} \quad B_2 \subseteq B_1.$$

The fourth step is to show that $\cup C$ is a maximal element of A. By the hypothesis of Zorn's Lemma $\cup C \in A$ since C is a chain which is a subset of A. Suppose $\cup C$ is not a maximal element, and let

$$G = \{D: D \in A \mathbin{\&} \cup C \subset D\}.$$

Let $D\star$ be the R-first element of G. Now for any $B \in C$

$$B \subseteq \cup C,$$

whence

$$B \subset D\star,$$

but then

$$\tau(f \mid \mathcal{S}(A,R,D\star)) = 1;$$

that is,

$$f(D\star) = 1$$

and

$$D\star \in C,$$

which is absurd. Thus our supposition is false and $\cup C$ is a maximal element as desired.   Q.E.D.

We now complete one cycle of implications by proving:

THEOREM 14. *Zorn's Lemma* Z_1 *implies the axiom of choice* AC_1.

PROOF. Since the kind of reasoning used in this proof is rather similar to the previous one, some details are omitted.

Let A be any non-empty set, and let

$G = \{g: (\exists B)(g$ is a function on B & $B \subseteq \wp A \sim \{0\}$ & for every non-empty set A_1 in B, $g(A_1) \in A_1)\}$.

G is, of course, the class of choice functions defined on some set of non-empty subsets of A. We use Zorn's Lemma to show that G must have as a member at least one function defined for all non-empty subsets of A.

Let C be any chain which is a subset of G. Then clearly $\cup C \in G$, whence Zorn's Lemma is applicable to G. Let f be a maximal element of G. I say:

$$\text{Domain of } f = \wp A \sim \{0\}.$$

Suppose not. Then there is a non-empty subset B_1 of A such that

$$B_1 \not\subseteq \mathfrak{D}f.$$

Let z_1 be some element of B_1, and define:

$$f_1 = f \cup \{\langle B_1, z_1 \rangle\}.$$

Obviously

$$f \subset f_1$$

and

$$f_1 \in G,$$

whence f is *not* a maximal element of G, which is absurd. Our supposition is false, and f is a choice function for A. Q.E.D.

We have now established the equivalences:

> *Axiom of choice* $AC_1 \leftrightarrow$ *numeration theorem*
> $\leftrightarrow$ *well-ordering theorem*
> $\leftrightarrow$ *Zorn's Lemma* Z_1.

We next want to show that the law of trichotomy implies one of the four statements just listed. (Theorem 12 together with the above equivalences establishes that any one of the four implies the law of trichotomy.) To this end, using some results of §7.3, we prove

THEOREM 15. *The law of trichotomy implies the numeration theorem.*

PROOF. Let A be any set. We want to show that A has the same power as some ordinal by using the law of trichotomy. Using Hartogs' function $\mathcal{H}$ introduced in §7.3, we have by virtue of Theorem 57 of §7.3

$$\text{not } \mathcal{H}(A) \leq A,$$

and thus by the law of trichotomy

(1) $A \prec \mathfrak{IC}(A).$

From (1) it follows immediately that there is a subset L of $\mathfrak{IC}(A)$ such that

(2) $A \approx L.$

Furthermore, since $\mathfrak{IC}(A)$ is an ordinal (Theorem 56 of §7.3), L is well-ordered and hence by the representation theorem for well-ordered sets (Theorem 81 of §7.4), there is an ordinal α such that

(3) $L \approx \alpha.$

From (2), (3), and the transitivity of equipollence the desired conclusion follows. Q.E.D.

We may now use the notion of a maximal chain to formulate Hausdorff's maximal principle [1914, pp. 140-41].

HAUSDORFF MAXIMAL PRINCIPLE: H_1. *If A is a family of sets, then every chain in A is a subset of some maximal chain in A.*

A somewhat simpler but equivalent formulation is:

H_2: *Every family of sets has at least one maximal chain.*

We state without proof:

THEOREM 16. H_1 *is equivalent to* Z_1.

THEOREM 17. H_2 *is equivalent to* H_1.

Formulations similar to Z_1, H_1, and H_2 are easily given for an arbitrary relation rather than inclusion.

DEFINITION 4.

   (i) *y is an R-upper bound of A if and only if for every x in A, $x \, R \, y$;*
   (ii) *y is an R-maximal element of A if and only if for every $x \neq y$ in A, not $y \, R \, x$.*
   (iii) *A is an R-chain of B if and only if $A \subseteq B$ and R simply orders A.*

Using these notions, we formulate:

ZORN'S LEMMA Z_2: *If $A \neq 0$, R partially orders A and every R-chain of A has an R-upper bound in A then A has an R-maximal element.*

We state without proof:

THEOREM 18. Z_2 *is equivalent to* Z_1.

Consideration of analogues for arbitrary relations of H_1 and H_2 is left as an exercise.

Finally we formulate a maximal principle due independently to Teich-müller [1939] and Tukey [1940].

DEFINITION 5. *A is a set of finite character if and only if*

(i) *A is a non-empty set of sets,*

(ii) *every finite subset of a member of A is also a member of A,*

(iii) *if every finite subset of a set is a member of A then the set is also a member of A.*

The intuitive idea behind this formulation is that a property is of finite character if a set has the property when and only when all its finite subsets have the property. A simple example of such a property is the following. Let R partially order A. Then R simply orders A if and only if R simply orders every finite subset of A. (In fact, it is sufficient to consider only the two element subsets of A.)

TEICHMÜLLER-TUKEY LEMMA: T. *Any set of finite character has a maximal element.*

We prove:

THEOREM 19. *The Teichmüller-Tukey Lemma T is equivalent to Zorn's Lemma Z_1.*

PROOF. The proof is only given in outline. We first show that Z_1 implies T. Let A be a set of finite character, and let C be any chain which is a subset of A. To apply Z_1 we need to prove that $\bigcup C \in A$. Let F be a finite subset of $\bigcup C$. Then F is a subset of the union of a finite collection D of members of C, for each element of F must belong to some member of C and there are only a finite number of elements in F. Now since D is finite and is a subset of the chain C, it has a largest member, say E; and F must be a subset of E, for otherwise C would not be a chain. $E \in A$, whence since A is a set of finite character, $F \in A$; but then also $\bigcup C \in A$. The hypothesis of Z_1 is thus satisfied by A and by virtue of Z_1, A has a maximal element.

We now show that T implies Z_1. It is most convenient to prove that T implies H_2, which by virtue of Theorems 16 and 17 is equivalent to Z_1. Let A be a family of sets. Define:

$$B = \{C: \ C \subseteq A \ \& \ C \text{ is a chain}\}.$$

As previous remarks have indicated, it is obvious that B is a set of finite character, whence B has a maximal element, say $C\star$, but $C\star$ is a maximal chain in A; for if it were not, it would not be a maximal element of B, and this establishes H_2. Q.E.D.

The theorems of this section and the preceding one have established the following equivalences:

Axiom of choice AC_1 $\leftrightarrow$ axiom of choice AC_2
$\leftrightarrow$ axiom of choice AC_3
$\leftrightarrow$ numeration theorem
$\leftrightarrow$ well-ordering theorem
$\leftrightarrow$ Zorn's Lemma Z_1
$\leftrightarrow$ law of trichotomy
$\leftrightarrow$ Hausdorff Maximal Principle H_1
$\leftrightarrow$ Hausdorff Maximal Principle H_2
$\leftrightarrow$ Zorn's Lemma Z_2
$\leftrightarrow$ Teichmüller-Tukey Lemma T.

A classical open problem of set theory is the independence of the axiom of choice (or its equivalents) with respect to the remaining axioms of Zermelo set theory. The likelihood of its being independent is very high. The source of philosophical interest in this problem is the non-constructive character of the axiom of choice. Adoption or rejection of this axiom has aroused perhaps more controversy among mathematicians in this century than any other single question in the foundations of mathematics. Mathematicians with constructive leanings object to postulating the existence of a choice function when no indication is given of how this function is constructed. (For a lively discussion of this point, see the Supplementary Notes to Borel [1950], and also Sierpinski [1928, Chapter 6].)

The claim that use of the axiom of choice may lead to a contradiction is definitely false in the following sense. It has been proved by Gödel ([1938], [1940]) that the axiom of choice is relatively consistent, that is, if the other axioms of set theory are consistent, the addition of this axiom will not lead to a contradiction.

On the other hand, application of the axiom of choice can lead to some paradoxical results. Perhaps the most celebrated example is the Banach-Tarski paradox [1924] that by using this axiom a sphere of fixed radius may be decomposed into a finite number of parts and put together again in such a way as to form two spheres with the given radius. More generally, Banach and Tarski showed that, in a Euclidean space of dimension three or more, two arbitrary bounded sets with interior points are equivalent by finite decomposition, that is, the two sets are able to be decomposed into the same finite number of disjoint parts with a 1–1 correspondence of congruence between their respective parts.

EXERCISES

1. Prove Theorem 10.
2. Prove Theorem 11.
3. Prove Theorem 12.

4. Consider the power set $\mathcal{P}\omega$ of the set of all integers. Give an example different from the one in the text of a maximal chain in $\mathcal{P}\omega$.

5. Prove Theorem 16.

6. Prove Theorem 17.

7. Prove Theorem 18.

8. Formulate a Hausdorff maximal principle which stands to H_1 as Z_2 stands to Z_1. Prove its equivalence to H_1.

9. Are the properties defined in Definitions 10-17 of §3.2 properties of finite character? (The method for reformulating them in terms of sets is obvious.)

10. Give an example of a property of a relation with respect to a set which is not of finite character.

§ 8.3 Axioms Which Imply the Axiom of Choice.

Without giving exact details we conclude this chapter with a brief discussion of two axioms which, together with the Zermelo-Fraenkel axioms listed in §7.5, imply the axiom of choice, but not necessarily conversely.

At the end of Chapter 6 we mentioned the Generalized Continuum Hypothesis, which asserts that for any infinite set A there is no set B such that $A < B < 2^A$. Lindenbaum and Tarski [1926] stated without proof that this hypothesis implies the axiom of choice; a proof of this fact has been published by Sierpinski [1947].

A very strong axiom which implies the axiom of choice as well as certain other axioms like the power set axiom is Tarski's axiom for inaccessible sets ([1938], [1939]). Before stating the axiom it will be useful to consider the problem which gave rise to the axiom, namely, the problem of the existence of inaccessible cardinal or ordinal numbers. We may characterize inaccessible cardinal numbers along the following lines. For each set A of cardinals we may show, on the basis of results in Chapter 7, that there is a smallest cardinal succeeding all members of A. This cardinal we may denote by sup A. Thus, $\aleph_0 = \sup \{1,2,3,\ldots,n,\ldots\}$, and $\aleph_1 = \sup \{\aleph_0\}$. A cardinal $\mathfrak{m}$ which is not 0 is said to be *inaccessible* if (i) for every set A of cardinals such that $\overline{\overline{A}} < \mathfrak{m}$ and $\mathfrak{n} < \mathfrak{m}$ for $\mathfrak{n}$ in A we have:

$$\sup A < \mathfrak{m}$$

and (ii) if $\mathfrak{n} < \mathfrak{m}$ and $\mathfrak{p} < \mathfrak{m}$ then $\mathfrak{n}^{\mathfrak{p}} < \mathfrak{m}$.

Obviously $\aleph_0$ is inaccessible in the sense of this definition. The fundamental question is: Can the existence of any other inaccessible cardinal numbers be established on the basis of the Zermelo-Fraenkel axioms, including the axiom of choice? It has, in fact, been shown by Shepherdson [1952] that the postulate that there are no other such inaccessible cardinals is consistent with the axioms of von Neumann-Bernays-Gödel set theory, and with little modification his proof holds for Zermelo-Fraenkel set theory.

It was for the purpose of establishing the existence of inaccessible numbers that Tarski introduced his *axiom for inaccessible sets:*[*]

For every set N there is a set M with the following properties:

  (i)   *N is equipollent to a subset of M*;
 (ii) $\{A: A \subseteq M \ \& \ A \prec M\}$ *is equipollent to M*;
(iii)   *there is no subset P such that the set of all subsets of P is equipollent to M.*

Tarski has shown that the cardinal number of a set M is infinite and inaccessible if and only if M satisfies conditions (ii) and (iii) of the axiom.

§8.4 Independence of the Axiom of Choice and the Generalized Continuum Hypothesis.

Since the first edition of this book was published, Paul J. Cohen has shown that the axiom of choice is independent of the other axioms of Zermelo-Fraenkel set theory. Gödel proved earlier that if Zermelo-Fraenkel's set theory is consistent, then it remains consistent if the axiom of choice and the generalized continuum hypothesis are added. Cohen also proved that the generalized continuum hypothesis is independent of Zermelo-Fraenkel set theory plus the axiom of choice. A good account of Cohen's results is to be found in his book, *Set Theory and the Continuum Hypothesis*, W. H. Benjamin, Inc., New York, 1966.

Using the methods developed by Cohen, a large number of additional important results on the foundations of set theory have been established; but it is beyond the scope of this book to survey them all. A quite detailed account of Gödel's original methods and the additional important methods developed by Cohen is to be found in a book by A. Mostowski, *Constructible Sets with Applications*, North-Holland Publishing Company, Amsterdam, 1969.

From a general philosophical standpoint Cohen's proof of the independence of the continuum hypothesis from the standard axioms of set theory is one of the most important results in the foundations of mathematics in several decades. To prove the continuum hypothesis was the first problem on Hilbert's famous list of open problems given at the International Congress of Mathematicians in 1900, and Hilbert assumed it could be proved.

The independence of the continuum hypothesis and many related assertions show the intuitive incompleteness of the standard axioms of set theory—incompleteness to a degree hardly expected. To a

---

[*]The version given here is that of Tarski [1939], which is an improvement of the one given in Tarski [1938].

certain extent the situation is comparable to that in geometry after the independence of the parallel postulate was established. We now have many geometries and presumably we may expect many set theories. However, the intuitive consequences for the foundations of mathematics are not certain. Will there be a turning away from set theory as the standard foundation of mathematics? The answer may be affirmative, but it is likely that the main lines of development of Zermelo-Fraenkel set theory will retain a permanent place nearly comparable to that given Euclidean geometry.

REFERENCES

Bachmann, H. *Transfinite Zahlen.* Berlin, 1955.

Banach, S., and A. Tarski. "Sur la décomposition des ensembles de points en parties respectivement congruentes," *Fundamenta Mathematicae,* Vol. 6 (1924), pp. 244-277.

Bernays, P. "A system of axiomatic set theory: I-VII," *Journal of Symbolic Logic,* Vol. 2 (1937), pp. 65-77; Vol. 6 (1941), pp. 1-17; Vol. 7 (1942), pp. 65-89, 133-145; Vol. 8 (1943), pp. 89-106; Vol. 13 (1948), pp. 65-79; Vol. 19 (1954), pp. 81-96.

Beth, E. W. *Les Fondements logiques des mathématiques.* Gauthier-Villars, Paris, 1950.

Bolzano, B. *Paradoxien des Unendlichen.* Leipzig, 1851.

Borel, E. *Leçons sur la théorie des fonctions,* 4th ed. Paris, 1950.

Burali-Forti, C. "Una questione sui numeri transfiniti," *Rendic. Palermo,* Vol. 11 (1897), pp. 154-164 and 260.

Cantor, G. *Contributions to the Founding of the Theory of Transfinite Numbers.* Translated by P. E. B. Jourdain. Chicago, 1915. Reprinted recently by Dover Publications, Inc.

Church, A. "The Richard paradox," *Amer. Math. Monthly,* Vol. 41 (1934), pp. 356-361.

——— "The constructive second number class," *Bull. Amer. Math. Soc.,* Vol. 44 (1938), pp. 224-232.

——— *Introduction to Mathematical Logic.* Princeton University Press, Princeton, 1956.

Church, A., and S. C. Kleene. "Formal definition in the theory of ordinal numbers," *Fundamenta Mathematicae,* Vol. 28 (1937), pp. 11-21.

Dedekind, R. *Was sind und was sollen die Zahlen?* Braunschweig, 1888.

Denjoy, A. *L'Enumeration transfinie: I. La notion de rang.* Paris, 1946.

Fraenkel, A. "Zu den Grundlagen der Cantor-Zermeloschen Mengenlehre," *Math. Annalen,* Vol 86 (1922a), pp. 230-237.

——— "Über den Begriff 'definit' und die Unabhängigkeit des Auswahlaxioms," *Sitzungsb. d. Preuss. Akad. d. Wiss., Physik. math. Klasse* (1922b), pp. 253-257.

——— *Einleitung in die Mengenlehre,* 3d ed. Berlin, 1928.

——— *Abstract Set Theory.* North-Holland Publishing Company, Amsterdam, 1953.

Frege, G. *Grundgesetze der Arithmetik,* Vols. I and II. Jena, 1893 and 1903.

Geach, P., and M. Black. *Translations from the Philosophical Writings of Gottlob Frege*. Blackwell, Oxford, 1952.

Gödel, K. "The consistency of the axiom of choice and of the generalized continuum hypothesis," *Proc. Nat'l Academy of Sciences*, U.S.A., Vol. 24 (1938), pp. 556-557.

—— *The Consistency of the Axiom of Choice and of the Generalized Continuum Hypothesis with the Axioms of Set Theory*. Annals of Mathematics Studies, No. 3, Princeton, 1940.

Grelling, K., and L. Nelson. "Bemerkungen zu den Paradoxien von Russell und Burali-Forti," *Abh. der Friesschen Schule*, N. S. Vol. 2 (1908), pp. 301-324.

Hartogs, F. "Über das Problem der Wohlordnung," *Math. Annalen*, Vol. 76 (1915), pp. 438-443.

Hausdorff, F. *Grundzüge der Mengenlehre*, 1st ed. Leipzig, 1914. Reprinted by Chelsea Publishing Company, New York, 1949.

Kelley, J. L. *General Topology*. Van Nostrand, Princeton, 1955.

Kuratowski, C. "Sur la notion de l'ensemble fini," *Fundamenta Mathematicae*, Vol. 1 (1920), pp. 129-131.

—— "Sur la notion de l'ordre dans la théorie des ensembles," *Fundamenta Mathematicae*, Vol. 2 (1921), pp. 161-171.

—— *Topologie*, Vol. I. Monogr. Mat., T. III, Varszawa and Lwow, 1933.

Landau, E. *Grundlagen der Analysis*. Leipzig, 1930. Reprinted, New York, 1946.

Lindenbaum, A., and A. Tarski. "Communication sur les recherches de la théorie des ensembles," *Comptes rendus Varsovie*, Vol. 19 (1926), pp. 299-330.

Mirimanoff, D. "Les antinomies de Russell et de Burali-Forti et le problème fondamental de la théorie des ensembles," *Enseignement mathématique*, Vol. 19 (1917), pp. 37-52.

Montague, R. "Zermelo-Fraenkel set theory is not a finite extension of Zermelo set theory," *Bull. Amer. Math. Soc.*, Vol. 62 (1956), p. 260 (Abstract).

Neumann, J. von. "Zur Einführung der transfiniten Zahlen," *Acta Szeged*, Vol. 1 (1923), pp. 199-208.

—— "Eine Axiomatisierung der Mengenlehre," *Journal für die reine und angewandte Mathematik*, Vol. 154 (1925), pp. 219-240.

—— "Zur Hilbertschen Beweistheorie," *Math. Zeitschrift*, Vol. 26 (1927), pp. 1-46.

—— "Die Axiomatisierung der Mengenlehre," *Math Zeitschrift*, Vol. 27 (1928a), pp. 669-752.

—— "Über die Definition durch transfinite Induktion und verwandte Fragen der allgemeinen Mengenlehre," *Math. Annalen*, Vol. 99 (1928b), pp. 373-391.

—— "Über eine Widerspruchsfreiheitsfrage in der axiomatischen Mengenlehre," *Journal für die reine und angewandte Mathematik*, Vol. 160 (1929), pp. 227-241.

Peirce, C. S. *Collected Papers*, Vols. II-IV. Edited by C. Hartshorne and P. Weiss. Cambridge, Mass., 1932.

Quine, W. V. *Mathematical Logic*. New York, 1940.

—— "On Frege's way out," *Mind*, Vol. 64 (1955), pp. 145-159.

Ramsey, F. P. "The foundations of mathematics," *Proc. London Math. Soc.*, (2) Vol. 25 (1926), pp. 338-384.

Richard, J. "Les principes de mathématiques et le problème des ensembles," *Revue gen. des sc.*, Vol. 16 (1905), p. 541.

Robinson, J. "Definability and decision problems in arithmetic," *Journal of Symbolic Logic*, Vol. 14 (1949), pp. 98-114.

Robinson, R. M. "The theory of classes. A modification of von Neumann's system," *Journal of Symbolic Logic*, Vol. 2 (1937), pp. 29-36.

Rosser, J. B. "The Burali-Forti paradox," *Journal of Symbolic Logic*, Vol. 7 (1942), pp. 1-17.

Russell, B. *Principles of Mathematics*. London, 1903.

——— *Introduction to Mathematical Philosophy*, 2d edition. London, 1920.

Scott, D. "Definitions by abstraction in axiomatic set theory," *Bull. Amer. Math. Soc.*, Vol. 61 (1955), p. 442 (Abstract).

Shepherdson, J. C. "Inner models for set theory: I-III," *Journal of Symbolic Logic*, Vol. 16 (1951), pp. 161-190; Vol. 17 (1952), pp. 225-237; Vol. 18 (1953), pp. 145-167.

Sierpinski, W. "L'axiome de M. Zermelo et son role dans la théorie des ensembles et l'analyse," *Bull. acad. sc. Cracovie*, (1918), pp. 97-152.

——— *Leçons sur les nombres transfinis*. Paris, 1928; 2d ed., 1950.

——— "L'hypothèse généralisée du continu et l'axiome du choix," *Fundamenta Mathematicae*, Vol. 34 (1947), pp. 1-5.

——— "Le dernier théorème de Fermat pour les nombres ordinaux," *Fundamenta Mathematicae*, Vol. 37 (1950), pp. 201-205.

——— *Hypothèse du continu*, 2d ed. New York, 1956.

——— *Cardinal and Ordinal Numbers*. Warsaw, 1958.

Skolem, T. "Einige Bemerkungen zur axiomatischen Begründung der Mengenlehre," *Wiss. Vorträge gehalten auf dem 5. Kongress der Skandinav. Mathematiken in Helsingfors* 1922 (published in 1923), pp. 217-232.

——— "Einige Bemerkungen zu der Abhandlung von E. Zermelo: 'Über die Definitheit in der Axiomatik'," *Fundamenta Mathematicae*, Vol. 15 (1930), pp. 337-341.

Slupecki, J. "St. Lesniewski's calculus of names," *Studia Logica*, Vol. 3 (1955), pp. 7-76.

Stäckel, P. "Zu H. Webers Elementarer Mengenlehre," *Jahresber. d. d. M.-V.*, Vol. 16 (1907), p. 425.

Suppes, P. *Introduction to Logic*. Van Nostrand, Princeton, 1957.

Tarski, A. "Sur quelques théorèmes qui équivalent à l'axiome du choix," *Fundamenta Mathematicae*, Vol. 5 (1924a), pp. 147-154.

——— "Sur les ensembles finis," *Fundamenta Mathematicae*, Vol. 6 (1924b), pp. 45-95.

——— "Über unerreichbare Kardinalzahlen," *Fundamenta Mathematicae*, Vol. 30 (1938), pp. 68-89.

——— "On well-ordered subsets of any set," *Fundamenta Mathematicae*, Vol. 32 (1939), pp. 176-183.

——— *Logic, Semantics, Metamathematics: Papers from 1923 to 1938.* Translated by J. H. Woodger. Oxford, 1956.

Teichmüller, O. "Braucht der Algebraiker das Auswahlaxiom?" *Deutsche Math.*, Vol. 4 (1939), pp. 567-577.

Tukey, J. W. *Convergence and uniformity in Topology.* Annals of Math. Studies, No. 2, Princeton, 1940.

Whitehead, A. N., and B. Russell. *Principia Mathematica*, 3 vols. Cambridge, England, 1910, 1912, 1913. 2d edition, 1925, 1927, 1927.

Wiener, N. "A simplification of the logic of relations," *Proc. of the Cambridge Philosophical Soc.*, Vol. 17 (1914), pp. 387-390.

Zermelo, E. "Beweis, dass jede Menge wohlgeordnet werden kann," *Math. Annalen*, Vol. 59 (1904), pp. 514-516.

——— "Untersuchungen über die Grundlagen der Mengenlehre: I," *Math. Annalen*, Vol. 65 (1908), pp. 261-281.

——— "Sur les ensembles finis et le principe de l'induction complète," *Acta Math.*, Vol. 32 (1909), pp. 185-193 (paper dated 1907).

——— "Über den Begriff der Definitheit in der Axiomatik," *Fundamenta Mathematicae*," Vol. 14 (1929), pp. 339-344.

——— "Über Grenzzahlen und Mengenbereiche," *Fundamenta Mathematicae*, Vol. 16 (1930), pp. 29-47.

Zorn, M. "A remark on method in transfinite algebra," *Bull. Amer. Math. Soc.*, Vol. 41 (1935), pp. 667-670.

GLOSSARY OF SYMBOLS

SYMBOL	NAME OF SYMBOL	PAGE
$-$	Negation	3
$\&$	Conjunction	3
$\vee$	Disjunction	3
$\rightarrow$	Implication	3
$\leftrightarrow$	Equivalence	3
$(\forall v)$	Universal quantifier	3
$(\exists v)$	Existential quantifier	3
$(E!v)$	Uniqueness quantifier	3
$=$	Identity	4, 14
$\in$	Set membership	6, 14
0	Empty set	14
$\notin$	Non-membership	21
$\subseteq$	Set inclusion	22
$\subset$	Proper inclusion	23
$A \cap B$	Intersection	25
$A \cup B$	Union	26
$A \sim B$	Difference	28
$\langle x,y \rangle$	Ordered pair	32
$\{x : \varphi(x)\}$	Definition by abstraction	34
$\cup A$	Union or sum of A	37, 44
$\cap A$	Intersection of A	39, 44
$\wp A$	Power set of A	47
$A \times B$	Cartesian product	49
$\mathcal{D} A$	Domain of A	59
$\mathcal{R} A$	Range of A	60
$\mathcal{F} A$	Field of A	61
$\breve{A}$	Converse of A	61
A/B	Relative product of A and B	63
$R \vert A$	R restricted to A	64
$R\,"A$	Image of A under R	65
$\mathcal{I} A$	Identity relation on A	70
$\mathcal{S}(A, R, x)$	R-segment of A generated by x	77
$R[x]$	R-equivalence class of x	81

SYMBOL	NAME OF SYMBOL	PAGE
$\mathbf{II}(R)$	Partition of A generated by R	84
$\mathbf{R}(\Pi)$	Relation generated by a partition	85
$f \circ g$	Composition of functions	87
A^B	Set of all functions from B to A	89
$\approx$	Equipollence	91
$\preceq$	Equal to or less pollent	94
$\prec$	Less power than or less pollent than	97
$\overline{\overline{A}}$	Cardinal number of set A	109, 242
$\mathcal{K}(A)$	Cardinal number of set A	111
$\mathfrak{m}'$	Successor of cardinal $\mathfrak{m}$	119
$\mathcal{Q}(\mathfrak{m})$	Set of all predecessors of cardinal $\mathfrak{m}$	120
$< \mid A$	Less than restricted to A	124
$\mathcal{E}A$	Membership relation on A	130
A'	Successor of set A	134
ω	Set of all finite ordinals	139
$\mathfrak{S}_A$	Successor function restricted to A	140
$\aleph_0$	Aleph null	156
$\lvert x \rvert$	Absolute value of x	172
$\mathfrak{c}$	Cardinal number of the continuum	193
$\{\alpha_\xi\}_{\xi < \mu}$	μ-sequence of ordinals	217

AUTHOR INDEX

Bachmann, H., 205
Banach, S., 250
Bernays, P., 12n, 241
Bernstein, F., 95n
Beth, E. W., 8
Black, M., 5n
Bolzano, B., 138n
Borel, E., 250
Brouwer, L. E. J., 1
Burali-Forti, C., 8-9, 133

Cantor, G., 1-2, 5, 9, 10, 91, 95n, 97, 109, 118, 127, 156, 159-61, 189, 191, 193, 242
Cauchy, A. L., 161n
Church, A., 10n, 14n, 229
Cohen, P. J., 252

Dedekind, R., 1, 99-100, 138, 159-60
Denjoy, A., 229

Epimenides, 9
Euclid, 1
Eudoxus, 1

Fermat, P., 217
Fraenkel, A., 8, 95n, 202
Frege, G., 1, 5n, 109, 111

Geach, P., 5n
Gödel, K., 12n, 194, 250
Goldbach, C., 217
Grelling, K., 10

Hartogs, F., 226, 247
Hausdorff, F., 245, 248

Jourdain, P. E. B., 2n

Kelley, J. L., 244n
Kleene, S., 229
Kronecker, L., 1
Kuratowski, C., 32, 64n, 99, 103, 245

Landau, E., 137n, 160
Lesniewski, S., 55
Lindenbaum, A., 251

Mirimanoff, D., 53n
Montague, R., 237
Moore, R. L., 245
Mostowski, A., 252

Nelson, L., 10
Neumann, J. von, 12n, 14n, 53n, 127, 129, 133n, 197, 202

Peano, G., 1, 121ff., 160
Peirce, C. S., 99n

Quine, W. V., 5n, 9, 55
Ramsey, F. P., 8, 139
Richard, J., 10
Riemann, G. F. B., 86
Robinson, J., 136
Robinson, R. M., 12n
Rosser, J. B., 9
Russell, B., 1, 5n, 6, 8-9, 30, 65n, 87n, 97, 99, 102, 109, 111, 138n, 139, 159-60, 171

Schröder, E., 95n
Scott, D., 127, 129
Shepherdson, J. C., 251
Sierpinski, W., 99, 103, 111, 128, 193-4, 205, 217, 229, 250, 251
Skolem, T., 8n, 202
Slupecki, J., 55
Stäckel, P., 108, 149

Tarski, A., 11n, 99-100, 107, 108, 111, 117, 122, 149, 250, 251-2
Teichmüller, O., 249
Tukey, J. W., 249

Weierstrass, K., 1, 86
Whitaker, J. M., 95n
Whitehead, A. N., 1, 65n, 87n, 102, 139, 171
Wiener, N., 32

Zermelo, E., 6, 7, 8n, 9, 10, 20, 53-5, 99, 139, 237, 240, 243
Zorn, M., 245

SUBJECT INDEX